GW00372465

Reinventing the Wheel

The Construction of **British Airways London Eye**
conceived and designed by Marks Barfield Architects

The original hardback edition of this book was published in July 2000
with the support of:

British Airways
Marks Barfield Architects
The Tussaud's Group

This revised and abridged softback edition was published in April 2002

Designed and edited by Ian Lambot

Printed and bound in Italy by Conti Tipocolor s. r. l.

ISBN   1 873200 31 5

Contents

As a friend of David Marks and Julia Barfield, I had known about their dream to build the world's largest observation wheel in the very heart of London from its earliest stages. Indeed, I had worked out of their offices for two years when I first returned from Hong Kong in the summer of 1995, and had watched from the sidelines, so to speak, as the dream grew ever closer to reality — first with the granting of the planning approval in 1996 and then with the formation of the partnership with British Airways.

Even then, I had my doubts whether the project would ever see the light of day. But Julia and, in particular, David's persistence, determination and sheer bloody-mindedness should never be underestimated. For various reasons, explained in part below, I have not covered this early side of the story in detail in this book, but David and Julia's success in bringing the project to the point where it really would be built cannot be underestimated. While recognising the importance of British Airways' and Bob Ayling's support throughout, it is a personal achievement every bit as great as the combined efforts of those involved since in bringing the project to its successful completion.

David and I had discussed the possibility of producing a book about the project's construction during those early days, but I can't say I had been that keen. In the 15 years since I published my first book under the Watermark imprint, all on architecture, engineering and design, I have produced several books on the construction of major building projects — most notably the new headquarters for Hongkong Bank in Hong Kong (1983-86) and the Commerzbank building in Frankfurt (1995-97) — so it was not that I had no experience of such an undertaking, it was more that I was unsure whether such a book would have broad appeal.

Foreword

In all of these earlier books I had concentrated on explaining what had been built, rather than why. In part, this comes out of my own preference for the craftsmanship of the building process. Also, and perhaps more importantly, these were buildings that had made significant advances in office planning and servicing; there were major lessons to be learnt that had direct relevance for architects of all persuasions. Surely a similar book about the construction of an observation wheel, even one of such unusual design, would only be of interest to the designers of other observation wheels — hardly a thriving industry. I needed a theme with wider implications.

As noted above, I have never felt inclined to explore how prestigious buildings such as these come about. I have observed, often at very close hand, the political intrigue that inevitably swirls around projects of this scale and complexity but, other than to those directly involved, I have never been convinced that this is of any specific long-term relevance. There is, certainly, much to be learnt about the general human condition in such accounts — the inevitable in-fighting, the need for points to be scored and scores to be settled — but that is a different book entirely, and one that I am happy to leave to others.

Understandably, therefore, I was not entirely sure what to say when David rang in December 1998 to say that, after many trials and tribulations, the project was at last going ahead. It was great news, but I was no closer to believing a book was viable. I was, however, sufficiently intrigued to attend a meeting a few days before Christmas to discuss the situation further. I was introduced to Tim Renwick and Fiona Hanan of Mace, and it was as they described how the project had developed and on what basis it was proceeding that the basic concept for this edition began to crystallise. As the meeting went on, I began to realise that what was of interest here was not so much the technology — though I have to admit I underestimated quite how daring much of this would be at that stage — rather it was the teamwork.

Listening to Tim explain the number of companies involved, their technical and geographical diversity, it became clear that success would only be possible if there was a true and complete understanding between all the parties involved — client, consultant and contractor alike. Everyone would have to play their part with a level of integrity, commitment and co-operation rarely seen on projects of this size. It was clearly going to be an extraordinary adventure and, quite frankly, I was hooked. I was not to be disappointed.

As the book developed through 1999 and I got to know many of those directly involved, the correctness of that initial analysis was only reinforced. In my experience, the level of

dedication, passion and co-operation invested in this project by everyone concerned was truly unique. And it is this, above all, that has made British Airways London Eye the success it is today.

It was a privilege to be part of that process and to spend time with many of those who gave so much to the project at every level of the design and construction process, from key designers and engineers to steel fabricators, from barge-masters and crane operators to systems specialists. I can only hope that by focussing on their efforts and, for the most part, letting them speak for themselves, I have captured something of their passion and inventiveness, their generosity and good will.

As I prepare this revised version of the book for reprinting — some 18 months after the publication of the original edition — the many kind comments I have received seem to indicate that I may, at least in part, have achieved some of my earlier aspirations. Certainly, going through the book again has brought back many happy memories. I can only hope that those approaching the book afresh have as much fun reading it as I had making it.

Many people helped me during the 18 months I spent researching, photographing, interviewing and generally recording the design and construction of this extraordinary structure. I cannot possibly name them all here, but to one and all I extend my sincere thanks.

Ian Lambot
London, March 2002

A big wheel for London

Peter Popham welcomes a bit of vulgarity on the river.

appointed to a second
_n of the Royal Fine
is dead against it.
Society wants to
_nk Art Centre's
_t the way it is:
_ir teeth over
_ng. All that's
_f Wales to
_gue – an
_om the
_ngler

But the grounds are in any case very
strong. As the architect Piers Gough
has pointed out, there has always been
a dichotomy between the dignities of the
Thames's North Bank – parliament, St
Ben, Somerset House – and on the South
the frivolity of the South B
art and theatre, its
trust was
and Oxo Tower
today

In recent tim
become
der

8

"There is an innate desire in all men to view the earth and its cities and plains from 'exceeding high places', since even the least imaginative can feel the pleasure of beholding some broad landscape spread out like a bright coloured carpet at their feet, and of looking down upon the world as though they scanned it with an eagle's eye. For just as the intellect experiences a special delight in being able to comprehend all the minute particulars of a subject under an associate whole, and to perceive the previous confusion of the diverse details assume the form and order of a perspicuous unity; so does the eye love to see the country or the town, which it normally knows only as a series of disjointed parts (as abstract fields, hills, rivers, parks, streets, gardens or churches), become all combined, like the coloured fragments of the kaleidoscope, into one harmonious and varied scene. With great cities, however, the desire to perceive the dense multitude of houses at one single glance, instead of by some thousand different views, and to observe the intricate network of the many thoroughfares brought into the compass of one large web, as it were, is a feeling strong on everyone, the wisest as well as the most frivolous; upon all, indeed, from the philosopher down to the idler about town."

Henry Mayhew
After an ascent in a hot-air balloon, 1882.

The earliest sketches of what is now known as British Airways London Eye first saw the light of day in December 1993, prepared by the architectural husband and wife team of David Marks and Julia Barfield in response to an 'ideas' competition for a monument to celebrate the turning of the millennium. Their inspiration was to provide a sweeping view of London from a new landmark at the very heart of the city. Their initial research led to a remarkable series of drawings of the city *(left)* — made by W. L. Wyllie and H. W. Brewer in the 1880s in response to a new craze for panoramas made from hot-air balloons — which captured the spirit of their proposal exactly. Unaware, at this stage, of the parallels with George Ferris's and Walter Bassett's work on Great Wheels at the end of the nineteenth century, their initial design — developed in association with Jane Wernick of Ove Arup & Partners — for a visually light but stately wheel was prepared from first principles.

"It all started on 23 October 1993, when I came across a piece by Richard Rogers in that week's Sunday Times *announcing an ideas competition for a landmark to celebrate the millennium. The submission date was 4 January 1994. We didn't do much about it initially, other than some vague discussions that it had to be different: not another staid monument, but something uplifting; something that would put a smile on people's faces; something they could participate in. The trouble was we had no real idea what that might be, until the thought struck me, out of the blue really, that it had to be a view of London, a really complete view that would take in the whole city."*

David Marks
Director, Marks Barfield Architects

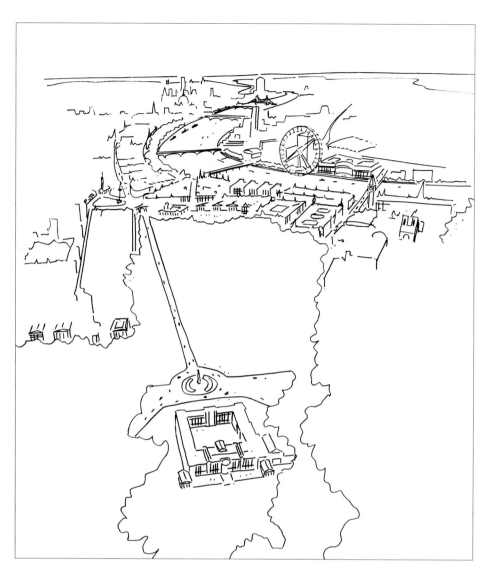

Drawn as part of the competition entry, these early sketches demonstrate clearly how little the original concept has changed over time. Indeed, a brief verbal description could easily be referring to the completed British Airways London Eye: the site is unchanged and the size is more or less the same; the capsules are held well away from the rim to allow unobstructed views in every direction; and the rim itself is supported on just one side and cantilevered out over the river. The only major difference was that in the search for greater visual lightness, these early designs explored the possibility of a single supporting 'arm' between the hub and the rim. In reality, this proved too heavy if sufficient 'stiffness' in the overall structure was to be maintained and later designs reverted to the use of more efficient — and hence far lighter — tensile spoke cables.

"I just knew it had to be Jubilee Gardens. If it was about London, it had to involve the river; and if it was about appreciating London from a new perspective, where better than from the heart of the city. Once you brought the two together, you soon realised there was no real alternative."

Julia Barfield
Director, Marks Barfield Architects

Nothing of substance was to come of the competition, but David and Julia's proposal captured many people's imaginations. With the encouragement of friends and relatives, they decided to press on and on 10 February 1994, together with David's father Mel, they set up The Millennium Wheel Company to develop and promote the concept. Aware of the controversy a new landmark on the skyline of London might provoke, a series of realistic photo-montages was prepared, showing how the wheel would look from various locations, including this view from the Houses of Parliament.

"We realised at the outset that consultation would be the key, and we spent a lot of time in those first few months arranging meetings – with planners from the London boroughs, with local MPs, with the landlords of County Hall, with the Port of London Authority, with the Arts Council and the South Bank Centre, and with all sorts of local interest groups. It was an interesting exercise. A lot of people grasped what we were trying to do immediately and were incredibly supportive, but others were not so sure. There was a definite element of elitism in many of the comments – that it would lower the tone, would attract hordes of ice-cream vans, that sort of thing – which is why we started doing more computer-montages, to prove that the wheel was above that, was something special. Then the Evening Standard picked it up and began a 'Back the Wheel' campaign which was incredibly helpful."

David Marks
Director, Marks Barfield Architects

13

By the end of 1994, David Marks and Julia Barfield had demonstrated that the building of the wheel was entirely feasible. It was clear, however, that significant investment was required if the project was to proceed further. Following a meeting shortly before Christmas with Bob Ayling, then British Airways Chief Executive, and a series of top-level discussions with the airline's finance and marketing team to evaluate the project, British Airways came on board as a partner in July 1995. Immediately, Jane Wernick and her team at Ove Arup & Partners were appointed to carry out detailed design studies necessary to prepare the project for full planning approval.

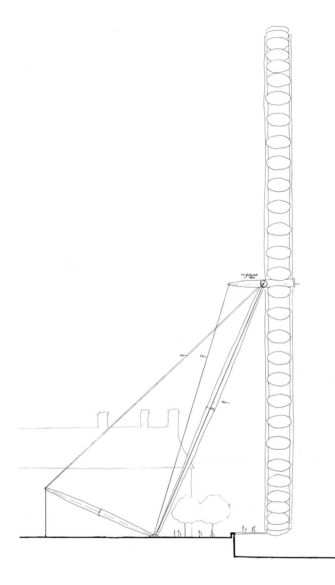

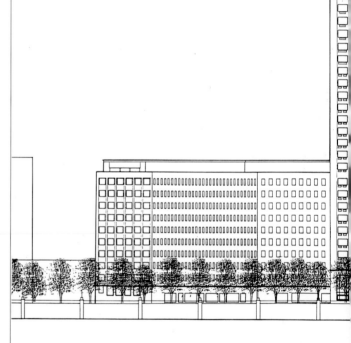

Early studies on the road to Ove Arup & Partners' final design *(far right)* included a sequence of drawings comparing the number of cables with the size of the rim *(right)* as well as sketches exploring how the wheel might be supported from as few points as possible *(above)*.

"I first became involved in the project in December 1993, when David and Julia asked if I would be willing to help with their competition entry. This was essentially in my spare time at that stage, with help from my colleague Sophie le Bourva. Then, it went quiet for most of '94 and I didn't really get involved again until the beginning of 1995. They were talking to British Airways by then and needed some additional advice on costs and so on. But the real design work only

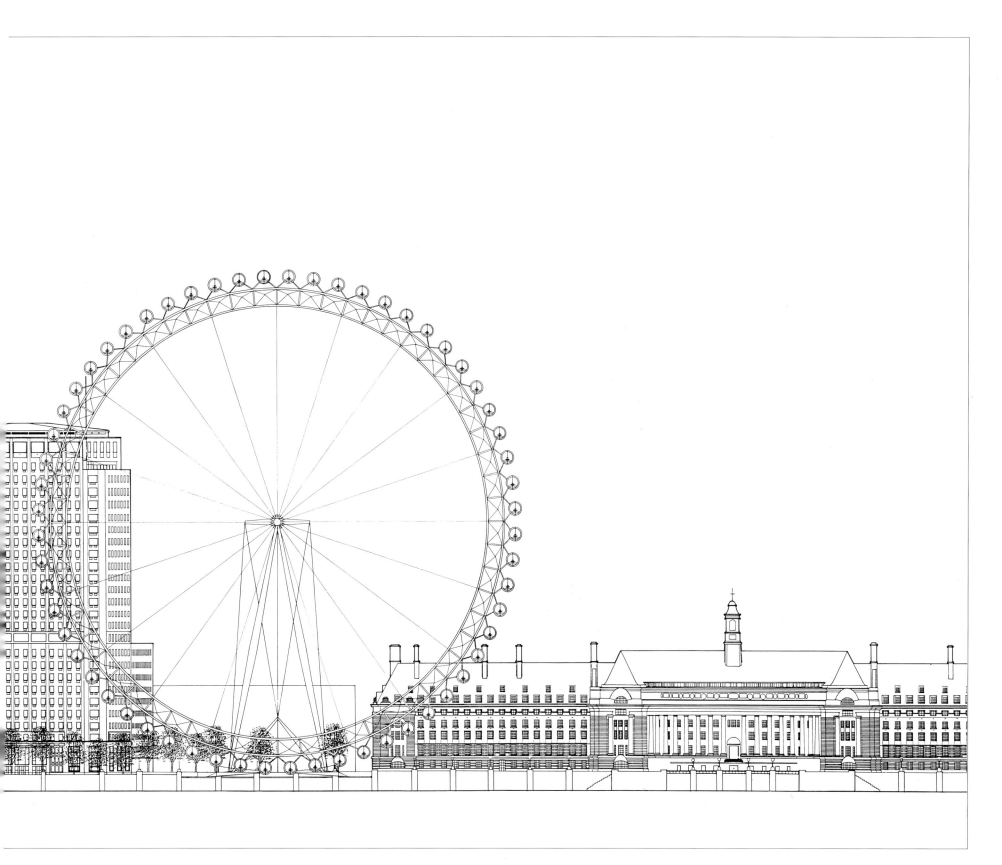

moved forward later in the year, when British Airways committed to underwrite the design development costs – around September I think. Then, everything started to happen very quickly. A whole series of studies were implemented: traffic movements, visitor predictions, buildabilty, power sources, capsule design, environmental controls, interference to radio transmission. The list was endless and it was my role to co-ordinate all the

information and feed it back into the design. We also undertook a full structural design review, which led to the use of spoke cables to support the rim rather than the single stiff arm of the competition entry. This, in turn, led to a very thorough series of studies looking at how the arrangement of cables and rim geometry affected issues such as metal fatigue and buckling of the rim. David and Julia were very concerned that the wheel should be visually as light as possible, so we made a series of drawings showing how

the number of cables affected the size of the rim. We settled on 20 pairs of cables, with each pair connecting one point on the rim's inner chord to either end of the hub."

Jane Wernick
Project Director, Ove Arup & Partners

"I first came on board in 1997, around July, when the project suddenly moved up a notch as it became increasingly clear it was really going to happen. London is British Airway's home base and the company brings more passengers, more travellers and overseas visitors into the UK than any other airline, so the project did have a business relevance, but it was clearly outside the norm. We consider ourselves experts on the technology of aeroplanes, for example, but the engineering here was entirely different – a lot of new technologies beyond our experience. This was also a tourist attraction, but while we might know a good deal about helping people to their destination, we don't normally become involved in what they do once they are there. At this stage, we still didn't have a complete financial package either, so it was all quite scary in many ways – tremendously exciting but also very fraught – trying to balance technical and commercial feasibilities and keep the project moving forward. But at the end of the day, you always knew you were working on something special, a project that was bound to make a big impact, and we all thought that it was just too exciting to be part of something so dramatic and let it pass by."

Paul Baxter
Project Director, British Airways London Eye

As part of the planning and consultation process, this beautiful model of London was prepared to prove that the wheel would not impinge unduly on the views from the city's other major landmarks. It served its purpose well and, on 22 October 1996, full planning consent was granted.

The design was now well advanced and a number of companies were approached to offer advice. The German company FAG prepared an initial design for the main bearings, while the Poma Group in France — specialists in the design fabrication of cable cars and ski-lifts — were commissioned to build a mock-up of the capsule. Babtie Allott & Lomax were appointed independent checking engineers, and the Dutch steel fabricator, Hollandia, advised on erection methods. By November 1997, the project was ready to go out to tender.

In November 1997 the project was put out to tender as a turn-key operation, with one company or consortium being asked to take full responsibility for the wheel's overall design and construction — a decision that was to mark the end of Ove Arup & Partners involvement in the project. The Japanese company Mitsubishi was selected in January 1998 and two months later Mace, one of the UK's leading project management firms, came on board to act as project manager. Two months later, in March 1998, The Tussaud's Group — who operate many of the UK's leading attractions and who had been considering the project for some time — also agreed to join the team, becoming a full partner in October.

As spring turned to summer, it became increasingly clear that Mitsubishi were struggling as the true complexity of the project became apparent. In June, the company proposed a compromise design, reducing the overall height of the wheel by 10 per cent to 135 metres and reducing the number of capsules from 60 to 32. With considerable reluctance, the proposal was accepted — a decision that was to form the basis of the final 'as-built' design shown here — but it was to no avail and in August 1998, by mutual consent, Mitsubishi withdrew from the project.

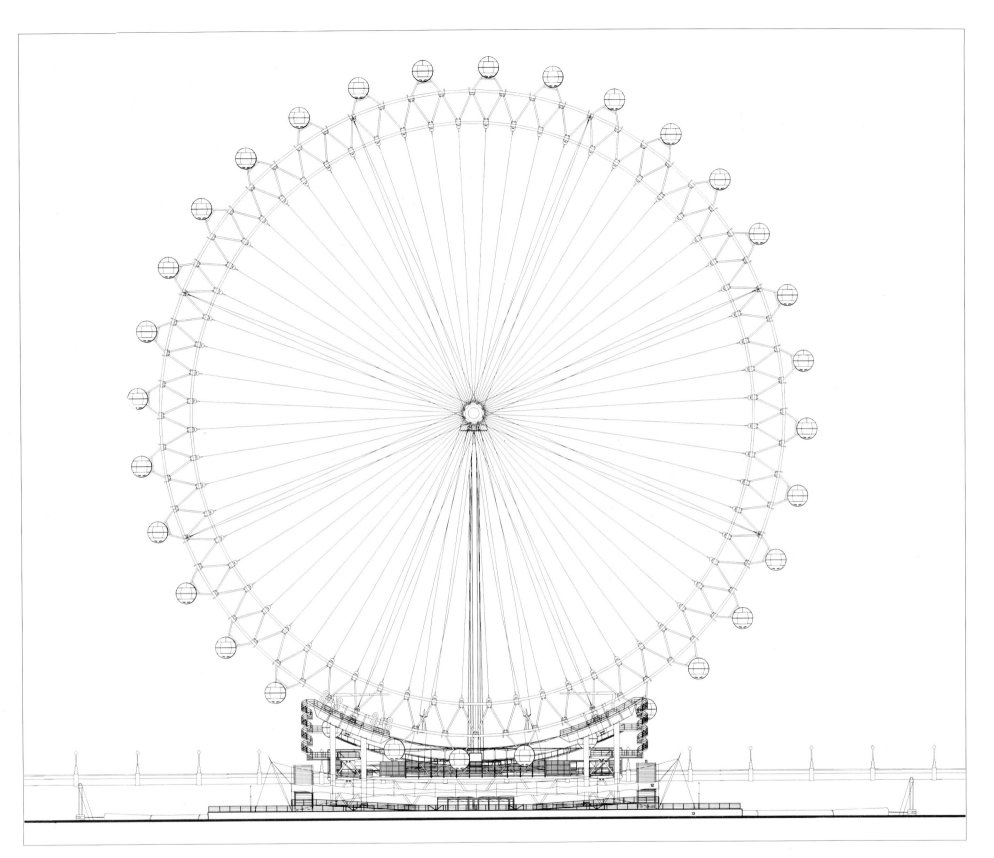

"When it became clear that Mitsubishi, for whatever reason, weren't going to be able to deliver the wheel everyone expected, it was a truly terrible moment. It seemed that was the end of it, but at the same time there was a real reluctance to let the project die. We had become too emotionally involved; we couldn't just walk away; we all knew we had to find a way. There seemed to be three alternatives: move ahead with a scheme none of us believed in; give up (and if we knew then just how difficult the project was going to be, we may well have advised that); or go back to Hollandia and Poma and see whether they felt able to take the job on, at short notice and nine months late, but still deliver on time. It seems mad that we even suggested it now, but somehow it set us all a challenge: if we went in this direction, and if Hollandia and Poma agreed, we were all accepting that we were going to do what-ever it took to complete the project. There would be no going back and no time for doubts. Hindsight is a wonderful thing but I think it's fair to say that everyone saying 'Yes, I'm in,' was one of the true defining moments of the project."

Tim Renwick
Project Director, Mace Limited

"In early December '97 I approached Marks Barfield Architects to ask if they realised there was a requirement under the Health & Safety Executive regulations to appoint an independent checking engineer for any project deemed to be a 'passenger carrying device'. Basically, this covers everything from roller-coasters to monorails, and has been put in place because most of these structures fall outside the scope of the standard building regulations. At its simplest, the independent checking engineer is responsible for anything to do with passenger safety, from assessing the initial engineering design, including methods and calculations, right through to physically testing the operating and safety systems. Everything has to be checked, tested and approved before a member of the public is admitted. We knew that very few architects or engineers would be familiar with these issues but, having worked on many major ride projects, we felt that the Babtie Group had a particular expertise to offer.

Following meetings with David and Julia, Jane Wernick and, of course, British Airways, we were appointed as the overall independent engineer for the project. We were soon faced with a dilemma: while it was essential we maintain our independent role, it was clear that this project was so complex it was in no one's interest to stand back and wait until the project was completed before passing judgment. Instead, we made the decision from the outset to become an

A computer-generated perspective *(left)* by Marks Barfield Architects, showing the final design prior to construction complete with all the major elements:

1 steel rim
2 hub and spindle
3 A-frame legs
4 spoke cables
5 main back-stay cables
6 capsule
7 boarding platform
8 restraint tower
9 pier
10 collision protection boom
11 compression base
12 tension base

In early September 1998, Hollandia formally agreed to join the project as contractor with responsibility for the engineering and construction of the wheel structure and its drive mechanism. Mace took on the role of construction manager and the Poma Group was appointed to design and build the capsules. The design team at Marks Barfield Architects *(below)* geared up for a new role overseeing the development of the final design — including work on the boarding platform, restraint towers, capsules and landscaping, the latter in association with landscape architect Edward Hutchison. The Sumitomo Bank and Westdeutsche Landesbank Girozentrale also came on board at this time, as part of a revised financial structure. In December, Tilbury Douglas Construction Ltd were awarded the contract for site and marine works. Following a formal groundbreaking ceremony on 28 January 1999 *(below left and bottom)*, British Airways London Eye was under way.

active member of the team, giving approval to the different design, fabrication and commissioning stages as they occurred. This came into its own when Hollandia and Poma came on board, as we were able to brief them impartially on all the design work that had been carried out previously."

Dr John Roberts
Project Director, Babtie Group

21

Following their formal appointment in September 1998 Hollandia undertook a complete review of the proposed design, carried out by their own engineers with support from a local consultant engineering firm they had worked with before, Iv-Infra. Aware that they did not have the capacity to complete all the work themselves in the time available, they also entered into negotiations with a rival steel fabricator, Mercon Steel Structures, to undertake the building of the rim sections. By late October, a clear direction had been determined allowing the structural design to move forward. The final detailed analysis would take many more months but it was now possible to prepare a complete set of shop drawings which allowed the first components to be put out to tender. Fabrication of certain parts began in a matter of weeks.

By mid–February, when the photographs shown here were taken, Hollandia had completed much of the detailed design work for the different elements that make up the main structure of the wheel, and production had begun in earnest in factories all over Europe. Many hundreds of dedicated craftsmen soon found themselves caught up in the excitement of working on this unique project. Shown here are just a few of the steelworkers who worked on the project at Mercon Steel Structures, their main efforts at this stage focussed on the fabrication of the complex tubular connections, or nodes, for the rim sections.

"Maarten (Jongejan) rang me out of the blue one afternoon asking if we could take on around 60,000 hours of work at short notice. We met a few days later and he told me what was going on and explained the work they had already done. I can't say I was too confident. It was an exciting challenge, but I don't think either of us was really sure if we would succeed."

Johan van Kuilenburg
Managing Director, Mercon Steel Structures BV

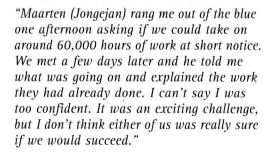

Cast-steel nodes were considered but the time required to complete all the necessary calculations before casting could begin would have added weeks to the programme. A delay at this stage could not be risked and it was decided, instead, to proceed on the basis of welded plate connections. Though not as smooth as castings, discussions between the architects and the engineers ensured a high visual quality was maintained.

"That first meeting at Rotterdam Airport came as a big surprise. As a contractor, you are hardly ever asked to work in a way that offers total control. The pull to say yes was very strong. Jobs of this profile and with these sorts of challenges do not come along very often, but this was not a decision you could make emotionally. I knew I had to be realistic, to tread carefully. I discussed the situation with my colleagues at Hollandia and made a few calls to people whose help we would need to make sure they had time available, then said I was interested but would need two weeks to analyse the drawings. Very quickly, it became clear there were a number of areas which needed to be looked at in more detail and we had to ask for more time. At this stage, we really could not be sure. We felt we were working at the limits of what was possible and any moment we might find something we could not control, so we began to pull in all the people we could think of to help: all of our own people, of course; specialist companies like Iv-Infra, Mercon, Skoda, Croese and FAG; and experts we knew with experience in cable design, damping systems, tube connections and so on — areas where we did not have too much experience. Always, we tried to find the best people in their fields. Only then did we have the confidence to move forward."

Maarten Jongejan
Managing Director, Hollandia BV

February 1999 and the master jig that will support the different components of the rim structure during final assembly takes shape in Mercon's main assembly hall *(left)*.

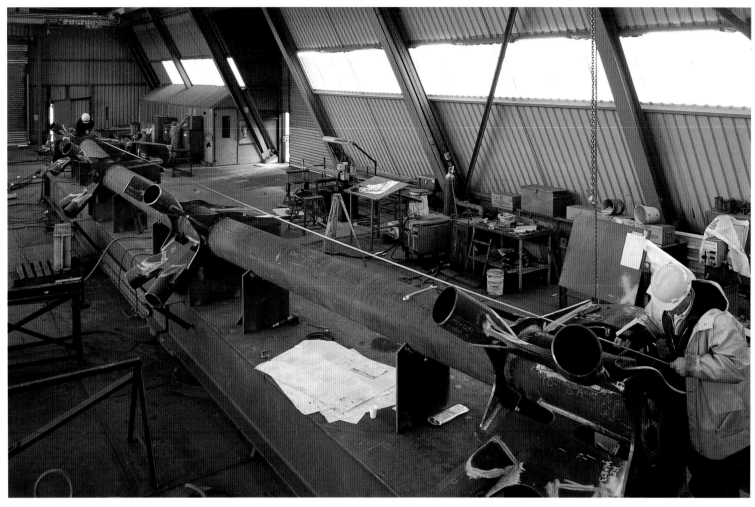

The fabrication of the rim sections began with the assembly of the smallest repetitive elements, the nodes. Hollandia had determined that direct tube to tube connections would not be strong enough to sustain the forces acting through these complex components and, working with TNO Bouw (the National Research Institute for Building Materials and Structures), had developed slotted connections using transverse and intermediate plates. These dramatically increase the contact area for each weld and improve the transfer of forces through the connection as a whole. Carefully considered welding sequences and purpose-designed jigs ensured the tight dimensional tolerances required were strictly maintained. As the nodes were completed, they were set out on ever larger jigs and joined by steel tubes to form the rim's inner chord *(above)* and then the outer frame *(following pages)* before the two were brought together in the main assembly hall.

The fabrication of the different pieces that make up the completed rim section was carefully broken down into discrete elements, steadily increasing in size and complexity but always in such a manner that overall dimensions and alignments could be adjusted and corrected at each stage to ensure the exacting tolerances were always met. The outer chords took shape in one part of the factory (bottom left) while the capsule support frames (right) were assembled in another. To ensure the pin-joint connections for the capsules were correctly aligned, Hollandia had prepared two matching jigs, one supplied to Mercon (below) and one to the capsule manufacturer, Sigma.

"We are specialists in welded and tubular-steel structures, so the fabrication of the individual parts of the rim was not that unusual. The tolerances were very small, so we had to look very carefully at all the welding sequences – first, how the different parts were made and then how all these parts were assembled – to make sure everything stayed as straight as possible. You must take care because every time steel is heated and cools it shrinks a little. But this is our usual work. Our real problem was that this was an aerial structure. With a building you start from a solid foundation and correct as you go, but with this there was no real reference point. Instead, we built a number of three-dimensional jigs in which we could position the pieces very accurately, but even then we had to be careful not to leave any stresses in the structure, which might have caused the section to twist later."

Johan van Kuilenburg
Managing Director, Mercon Steel Structures BV

Working from a master set of shop drawings supplied by Iv-Infra, the in-house fabrication drawings, welding and cutting schedules and other ancillary design work — including the design of the jigs — was carried out in the Mercon drawing office by the company's own engineers.

"There was always so little time. It was a bit like an express train: slow to start but difficult to stop once it was moving. But you also knew you were working on something special, something the whole world would know about. Even my neighbours, once they found out what I was working on, would stop me to ask how things were going which never happens usually."

Alex Rietveld
Project Engineer, Mercon Steel Structures BV

As the individual elements that make up the rim sections grew in size, ever greater care was necessary to ensure fabrication was carried out to the levels of accuracy required. Working up from smaller to larger sections allowed minor discrepancies to be corrected at each stage, but only within very fixed limits which became ever harder to achieve as the pieces grew in complexity. And the tolerances were very tight indeed. The 9-metre overall width of the outer frame for example, shown here during final assembly, could be exceeded by no more than 5mm. Of greater concern was the straightness of the whole frame. It was essential that movements out of plane in the completed rim, both horizontally and vertically, were kept to the absolute minimum to ensure a constant alignment with the boarding platform when the wheel was completed. Guiding systems were to be installed in the restraint towers to position the wheel correctly as it turned, but the greater the pressure required to keep the rim in place, the greater the fatigue loadings produced — both within the rim itself and in the restraint towers.

Work procedures were carefully planned with the welding teams and monitored throughout by Mercon's production managers to ensure the very highest quality was maintained at all times.

"Iv-Infra was involved from the very first day, helping with the analysis of the original proposal, then working closely with Professor Berenbak at Hollandia to develop their new design. Our main responsibility was to carry out the structural calculations and prepare all the drawings for the many steel components – the rim, hub, spindle, legs, restraint towers and boarding platform. Everything was interrelated, so we had to keep a good overview, but each element also had its own governing factors. The cables had to be pre-tensioned to avoid them going slack in certain wind conditions, but the greater the tension the greater the risk of buckling in the rim. There was also the problem of fatigue in the cables, the nodes and in the capsule support frames, caused by the turning of the wheel. Most of these were non-linear calculations which required the use of finite element analysis. In fact, the calculations were so complex and so critical that Hollandia asked the Municipal Engineering Office of the City of Rotterdam to check the calculations using a different computer program. There was also the problem of analysing movement and accelerations in the wheel due to varying wind loadings. Many of these calculations had to be repeated over time, with thousands of calculations for many points on the wheel. We had to consider every situation and the number of calculations was unique."

Arie Lanser
Project Manager, Iv-Infra

The assembly of the rim's inner chords and outer frames commenced at the beginning of March, and before the month was out the first of these were being brought together in the main assembly hall (main spread). Here, sections of the rim's triangular truss could be completed, three at a time, on a master jig to ensure that each section aligned perfectly with its neighbour. As each section was finished it was lifted, turned and placed into a separate jig, to allow its capsule support frames to be added to the 'top' of the outer frame. At this stage, the rim was fabricated in equal one-sixteenth sections. Measuring some nine metres wide and just over six metres deep (8.5 metres including the capsule support frame), each of these 24-metre long sections weighed 36 tonnes when complete and incorporated over 450 separate welded connections.

The final stage of the assembly of the rim sections, the fitting of the diagonal bracing between the inner chord and outer frame *(below)*, had to be carried out with complete accuracy. Great care was taken to position the different elements within the main jig precisely before welding began, and the welding procedures themselves were closely monitored. The slightest misjudgement might have induced twisting in the rim, which would have been very difficult to remove. Everything went smoothly and by April 1999 the first batch of completed one-sixteenth rim sections *(right)* were lined up outside Mercon's main assembly hall, ready for transportation by barge to Hollandia for shot-blasting, painting and final assembly.

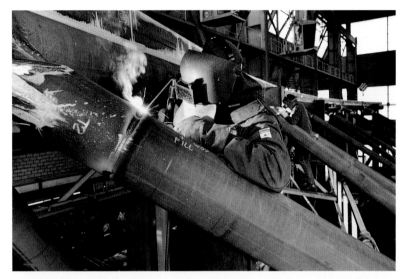

"It was very important that we minimised the building tolerances at every stage. This was relatively easy when the sections were in the factory, but much more difficult outside. We decided the best solution was to make a full three-dimensional survey of each section as it was completed so we knew its exact size and shape, which allowed us to set them out precisely before they were welded together."

Nardo Hoogendijk
Design Engineer, Hollandia BV

"My main impression of those first few months is that we had meetings rather than correspondence. I set up what were meant to be short progress meetings every Friday, but every week they turned into a major design review with long technical discussions. It was one of the few times when we were all together and we all understood that a problem for one of us was a problem for everyone. It was better to solve it there and then. Everything happened very fast – decisions were made and we moved on. If we were unsure about anything we would bring in an expert the following week and work with their advice. Always there was someone looking over our shoulder, not to take responsibility but to ask questions, make us think again, think carefully. We never looked back. It was a case of trying to identify what information could be released even though everything else was still in progress. That's how we were able to order the steel tube before we had completed the engineering drawings, and later to start the cutting schedules before we knew all the welding procedures. It was incredibly demanding, but also very exciting. And it was very successful – we had to change very little when the final calculations came through, and all the steel we ordered was used."

Chiel Smits
Project Manager, Hollandia BV/Mercon Steel Structures BV

On their arrival at Hollandia on the outskirts of Rotterdam, just a few hours by barge from the Mercon works in Gorinchem, the rim sections were lifted on to the quayside by floating crane and transferred as quickly as possible, first into Hollandia's purpose-built shot-blasting facility and then, when all the initial preparations had been completed, into the adjacent paint shop. Each hall is fitted with its own air-conditioning system to ensure low humidity levels are maintained at all times, thereby offering the very best conditions in which to carry out the cleaning, repairing and testing of the sections — all in the shot-blasting hall — as well as the final painting. Three coats of paint were applied: a zinc primer, followed by two coats of marine-quality white gloss epoxy to all parts of the rim except the drive rails along the sides of the outer frame, these being finished with a 'high-grip' aluminium magnesium coating.

In early May, just three short months after fabrication of the first node connections had begun, the first one-sixteenth rim sections were towed out of the paint shop and transferred to the main assembly areas – either at Hollandia (below) or Mercon (right) – ready to be welded together to form the one-quarter rim sections that would be shipped to London. Using the accurate surveys undertaken earlier, the completed one-sixteenth sections were first laid out very precisely to minimise any misalignments. The final connecting pieces – including a complete section of the inner chord, two diagonal braces and two short sections of the outer chords – could then be accurately cut to length and welded or, in the case of the inner chord, bolted into place. The final shot-blasting, weld testing and painting was then completed under full-height shrouding positioned where required.

Measuring some 88 metres long in their completed form, the one-quarter rim sections were larger than the barges that would be used to transport them to London by some margin. Careful preparations were required therefore, in particular with the setting out of the main support cradles, to ensure the rim was securely held in place and the whole barge was well balanced.

"Babtie Allott & Lomax have been involved in many major projects over the years, from large naval facilities and power stations to work on The Royal Opera House. We have also been involved with other 'rides', roller coasters and so on, both as checking engineers and as designers in our own right — and, set against these other projects, I can honestly say that this is one of the most complicated structural engineering projects we have ever undertaken. There were all the problems of fatigue analysis, of course, then the dynamic behaviour of the wheel and its response under different wind conditions. The use of heavy steel castings introduced a number of problems you don't usually come across, as did the whole issue of buckling in the rim due to the tension in the spoke cables. A great deal of analysis was required, but the real problem was understanding what sort of analysis and how to apply it. It can be quite difficult to get your mind round what is actually happening with a lot of these problems — when buckling occurs for example — but you need a fairly clear understanding if you want to know what to do about it. Not surprisingly, the project has made huge demands, both emotionally and time-wise, but it has also been tremendously exhilarating. I doubt whether many of us would have missed it for the world."

Allan Mann
Technical Director, Babtie Group

The end of May and the first of the completed one-quarter rim sections is being lifted on to the barge that will transport it across the North Sea and up the River Thames. The decision to move the rim in quarter sections had been determined by a number of factors, including what size cranes might be available on site and the difficulties that might be encountered when transporting the finished sections upriver. Detailed research had been undertaken, for example, to ensure the loaded barges would fit under all the bridges downstream from the site. At the same time, it was preferable to move the rim in as few sections as possible, as this minimised the need for time-consuming welding on site, as well as limiting the number of temporary platforms required in the river. To reduce the amount of on-site work to an absolute minimum, the rim sections were shipped as complete as possible, with their emergency access ladders, temporary work platforms and electrical connection rails already in place.

The main elements of the hub and spindle were, for the most part, manufactured in cast steel. Measuring 24.6 metres long and with an overall diameter of 2.15 metres at its widest point, the central spindle was far too large to cast as a single piece. Instead, it was produced in eight smaller sections, seven of which were cast while one, the straightforward cylindrical element between the bearings, was formed in rolled steel tube — this being more economical for the thickness of steel (80mm) required at that point. Two further castings, in the form of great rings some 4.6 metres in diameter and 900mm thick, form the main structural elements of the hub, a simple rolled steel tube forming the spacer that holds them apart. As with all the structural components, the design was carried out by Hollandia's in-house design team, with extra structural analysis from Iv-Infra. The casting work itself was carried out by Skoda Steel, its foundry at Plzen in the Czech Republic being one of the few capable of handling the large and very precise castings required.

The casting process begins with the fabrication of a full-size wooden pattern of the piece to be made, at the size it will be cast — usually a good deal larger than its final form as only the metal at the centre of the casting retains the high quality required. Shown here, during the preliminary stages, are the patterns for the two massive cast-steel 'feet' that support the main A-frame legs, but the same process was used for all the cast spindle sections, as well as the two cast components that make up the hub. To allow the work to proceed as quickly as possible, the detailed sizes of the spindle sections were released in stages, the external shape being fixed a week or so before the internal profile had been finalised.

"The forces acting within the hub are so complex that they always had to be castings. We did consider fabricating the spindle to save time, but this would have meant a much larger diameter and, consequently, unusually large bearings which would require a long lead-time as there would be a lot of design work to complete before fabrication could begin. We decided to talk to FAG in Germany — we had worked with them before — and it turned out they had already done a bearing design for Ove Arup & Partners based on a slimmer cast-steel spindle. This needed revising somewhat, but was basically ready to go straight into production. Because of the time factor, their design became the fix for the overall diameter of the spindle and that, in turn, meant it had to be a casting. Technically, we knew this was a good solution and, fortunately, it turned out Skoda were able to start work right away."

Professor Jacques Berenbak
Senior Design Engineer, Hollandia BV

The moulds are formed in a fine, densely packed black sand, mixed with a special resin to help it hold its shape. This is built up around the patterns in deep pits set into the casting hall's floor, with care being taken to ensure that the pattern, which may be built in sections, can be removed as the mould takes shape. Separate supply tubes set into the surrounding sand channel the molten steel to the right parts of the mould during the pour.

"Everything had to happen very quickly and there was no time for a formal tender process. Instead, we had informal discussions with the foundries we knew could do the work — to find out who was available, what they might charge, that sort of thing — and this identified Skoda very quickly. They had worked with us before, so we knew they could produce the quality we wanted and, because they knew us, they were willing to take a flexible attitude, to start the work before we had completed the design. We knew by then that if we wanted to keep to the schedule we had to cast the main pieces before Christmas so that we could use the three week holiday for the cooling down period. This left just 12 weeks for us to complete all the design work and for Skoda to make the patterns and moulds. It was very tight."

Peter Koorevaar
Technical Director, Hollandia BV

Although the casting of the main hub and spindle sections had been completed in January, the decision to assemble the wheel horizontally resulted in two further castings being required, both of which were poured on 6 April 1999. These were the two massive 'feet' that support the main A-frame legs. Weighing around 23 tonnes each in their final form (nearer 50 tonnes at the time of casting), these were the most difficult pieces to cast, as the combination of a very heavy central mass with relatively thin outer flanges made it more difficult to control the cooling of the different parts at a steady rate. The pouring of the molten steel — from the furnace into the crucible *(below)* and from the crucible into the mould *(right)* — was as dramatic as ever, particularly as at Skoda much of this work is still carried out by hand.

To ensure consistent strength throughout the thickness of the final casting, the speed with which the molten metal cools in the mould is as important a part of the process as the quality of the raw steel. Only a slow and even cooling will ensure the metallurgical properties of the steel are consistent from the centre to the edge. For the largest sections, this cooling might take up to four weeks, and is carefully monitored at all times by numerous sensors and heating elements built into the mould to ensure the temperature falls at the correct rate.

One of the hub castings, shown here being cleaned of unwanted material after the first of three heat treatments it will receive before final machining. These eliminate undue stresses in the metal and take place after each major stage in the machining process. At this stage, the casting weighed around 50 tonnes and was still considerably oversized — even if already much reduced by the removal of a riser weighing a further 30 tonnes. Though cast as a single element, each hub has two distinct layers: a large diameter circular flange, on to which are secured the 32 spoke cables, and, slightly above, a smaller faceted flange for the eight rotation cables. The machining itself was carried out in stages, beginning with the removal of the spigots along the sides of the casting — indicating where the molten steel entered the mould — and an initial rough machining. Three further machinings would be required before completion, by which time each hub section would measure 4.64 metres in diameter and weigh no more than 33.5 tonnes.

For ease of handling and to allow use of the most accurate turning machines, most of the spindle was cast in sections between two and three metres high. Shown opposite is one of the tapered sections that will form the back of the spindle during its second machining. Both the inside and outside of the section were machined, the entire operation being controlled by computer to ensure the correct wall thicknesses and angle of taper were achieved. With greater continuous strength required around the main pin-joint connection, only the central section was cast at an increased height. This is shown here after its first machining *(below)* – complete with an extra bulge where the bearing will sit to ensure the steel is of the highest quality at this critical point — and later *(right)*, during the second machining of the main pin-joint.

"I would not say any of the casting was a problem. We worked with Hollandia to decide how the sections were broken down, working out where we needed continuous strength and where welds were possible. We also had to consider the final machining, what was the best size for our turning machines. The most difficult piece was the centre part of the spindle – where there are the connections for the legs – but this was mainly because of its size. It weighed more than 180 tonnes when cast and was over seven metres high, which required a very big mould and careful control of where the metal entered, to ensure it was filled evenly and completely. The design of the mould comes partly from experience, but we also have our own computer programmes, developed here, which indicate which areas are critical so we can make the right allowances. We had to add a lot of extra material to the thickness of the main cylinder itself – with even more over the critical areas – and also to the height. This part is called the 'riser' and it is needed to squeeze all the flaws out from the centre of the casting. It was cut off as soon as the casting came out of the mould. When it was finished, the casting was five metres high and weighed around 70 tonnes, so you can see there is a big difference."

Václav Bocker
Sales Department, Skoda Steel

The most important of the hub and spindle sections were finished on specialist turning machines capable of working down to the 15-micron (0.015mm) tolerances required where the bearings are seated.

"The most demanding aspect of the project was the lack of time. This was particularly true of the spindle. There was no time to make any of the pieces again if we found a fault in the first casting, and this created a lot of pressure. Fortunately, I had worked with Skoda before so I knew they had the technical expertise. I had a good relationship, too, with most of the people there, which was a big help. Even so, I was visiting Plsen nearly every week in the first few months. But it all went very well. In fact, the only real mistake was mine. We had decided to cast a small plaque listing the main companies involved in the design and construction of the project on one of the feet, but I forgot to include Mace. Of course, they picked it up during a visit to Skoda, which was very embarrassing. Fortunately, we were able to weld the name on afterwards and now you cannot tell."

Adry Zondevan
Quality Assurance Manager, Hollandia BV

The final machining of the spindle and the two main parts of the hub was carried out in the Nuclear Machine hall, a separate section of the Skoda works usually dedicated to the production of complex components for power stations, nuclear reactors and the aerospace industries. Here, the different parts of the spindle were welded together to form three intermediate sections, two around 10 metres in length, this being the largest size that could readily be transported by road to Hollandia. The largest of these was the central section *(above)*, shown here having its temporary external fixings being removed after welding. Meanwhile, in another part of the workshop *(opposite)*, one of the hub sections is shown during final machining.

"The most important thing about this project was that we had to learn new ways of working with other people. The work itself was not that unusual: this was not the biggest casting we have ever made, and it was not the most challenging in the technological sense. What was special was that we had to work closely with the designers – receiving the information in a raw state, working from very simple drawings, making changes. I am not sure every foundry would have been able to work in such a way. We need a high degree of trust between ourselves and Hollandia, a good co-operation, a good understanding. The information came in stages, which required more communication and more discussion, and not everyone here was happy with this – the late changes and so on. Everything was very fast. Everything was different – the work, the agreements, the financial side. We had never worked like this and we had to find many new procedures. Fortunately, we had worked with Hollandia before and we have a good relationship with them, and that allowed us to work this way. It was not always easy for us, but it was a very good experience."

Václav Sístek
Deputy Managing Director, Skoda Trading

The massive A-frame legs that support the entire wheel were fabricated in tubular rolled steel by SIF, a Dutch company specialising in the making of pipelines and pressure vessels for the off-shore and petrochemical industries. Measuring some 58 metres in length and 3 metres in diameter at their centre, tapering down at each end, the two legs were delivered to Hollandia in their raw unpainted state, each leg in two sections. Hollandia had considered fabricating the legs themselves in thinner steel, lined internally with stiffening ribs, but quickly realised the time this would take was not cost effective, and elected instead to have the legs produced in 40mm-thick rolled steel. At 48 metres, the longer sections (below) were the maximum size that could be readily accommodated in the paint hall, while the shorter sections (right), the last 10 metres of the lower ends, were easier to handle during their final fit-out – a process which included the installation of the access hatches and the welding of the external fixing brackets into which are anchored the back-stay cables that hold the wheel vertical.

Due to the scale of the project and the speed of construction, it had been necessary for Hollandia to subcontract out a good deal of the initial steel fabrication. For a company which prides itself as much on the quality of its workmanship as on its organisational abilities, this was something of a disappointment. To make up for it, however, Hollandia did retain fabrication of one of the most complex elements: the connection at the top of the A-frame legs. The focus of massive forces, both in its final state and during the uplift, this integrates and links a total of three pin-joint assemblies: the permanent support for the hub and the spindle; the smaller anchor point for the top of the main backstay cables; and the anchor point for the cables used to lift the entire structure vertical. The use of very thick steel plates, combined with the complete accuracy required during their fabrication and assembly, transformed what might otherwise have been a purely practical element into something approaching a work of art.

"The initial assumption had always been that the wheel would be built vertically; this is what we had analysed during our work in 1997 for Arups, and it was our starting point when we began work on the project again in September '98. Very quickly, however, we realised there were real problems with the programme. It would just take too long, because you had to erect the legs and the hub and spindle before you could start the assembly of the rim. And the assembly

itself would be far more time-consuming: controlling the dimensional tolerances would be much more difficult and there were all the problems of working at height. By a strange coincidence, we had just started work on another job, for Mammoet, making special sections for the jib of a lifting gantry capable of lifting up to 3000 tonnes – much heavier than the wheel – and this started us thinking. Building the wheel horizontally brought

many advantages: the tolerances were far easier to control; it was safer; and it would be much quicker – we could meet the schedule. But there were problems too. We didn't know if we could get planning permission to build over the river; there were financial considerations; and it would mean a total reassessment of the design work we had done so far – we did not know what effect the lift would have on the structure. As it turned out, the rim needed very little new engineering, beyond calculations for

extra support cables to be used during the lift itself. The real impact was on the foundations – they had to be much bigger – and on the legs, particularly the connection at the head of the A-frame where all the lifting forces were concentrated."

Maarten Jongejan
Managing Director, Hollandia BV

The full 1000-tonne load generated by the rim, spoke cables and capsules is transferred to the spindle via two spherical roller bearings, purpose-made for the project by FAG OEM und Handel at its factories at Schweinfurt and Wuppertal in Germany. With inner core diameters of around 2.1 metres, the bearings were more-or-less standard items, the only unusual aspect being the use of drilled rolling elements held in place by a 'cage', rather than allowed to run free. This allows the rollers to be tightly spaced, helping to spread the high loads more evenly.

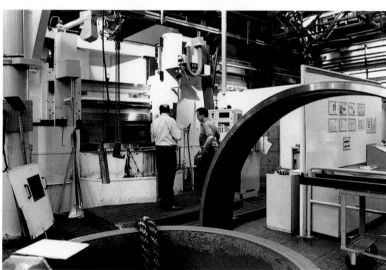

"We first became involved in this project as long ago as September 1995, when we were approached by Ove Arup & Partners to discuss the relative advantages of different types and sizes of bearings. They already had some ideas for the size of the spindle and we were able to prepare a design for them – a good solution. Then time went by; we transferred the design to FAG Japan but nothing came back, so it was a real surprise when Hollandia came to us with the same project. Fortunately, they were willing to base their design on our dimensions and only a few revisions were required, which allowed us to complete the project very quickly. Of course, every bearing of this size is unique to some degree, but there are set solutions to specific problems, so there is usually a precedent you can refer to. In fact, FAG made some bearings of a similar size around 20 years ago, so we can be confident of this design."

Gerhard Kleine
Production Manager, FAG OEM und Handel AG

"FAG has been making bearings for more than 100 years, so design now is more a case of evolution than new ideas – improving manufacturing techniques to give the bearings a longer life and so on. It depends on the maintenance, of course, and the general working conditions, but we would expect these bearings to last 50 years with no problems at all. We have bearings far older than that, still in operation"

Klaus Ptácnik
Quality Assurance Manager, FAG OEM und Handel AG

The final assembly of the bearings *(below left)*, each weighing around 6.2 tonnes, was completed at FAG's factory in Wuppertal, with the final fitting to the hub castings being carried out at Hollandia *(left and below right)*. Produced in case-hardened steel, the bearing's rolling elements have a gentle convex curve, the radius of which is slightly tighter than the matching concave curve of the raceways, to ensure the load is transferred through a narrow 'contact ellipse' well within the roller's overall length at all times. As might be expected, the tolerances are very small indeed: a mere 2 microns, or 0.002 mm, covering all the rolling elements, and around 10 microns for the raceways. Designed with a theoretical lifetime of 200,000 hours operational use — or 50 years, running at the present schedule — and considered all but fail-proof, it was nevertheless decided that allowance should be made to remove one or both bearings while *in situ*, just in case. The inner bearing was therefore made slightly smaller than its outer counterpart, to ensure there was adequate all round clearance should it ever need to be withdrawn past the outer bearing point.

It had been realised at the outset that the only real way to be entirely sure that the spindle was up to the job was to carry out a full-load test before it was transferred to site. With a total imposed load in excess of 1150 tonnes, this was easier said than done. Fortunately, Hollandia had a small assembly hall of just the right size *(right)*, and this was transformed into a full-size test rig. A massive, purpose-made tubular steel frame, complete with fixing straps, held the centre and rear of the spindle in place, while further straps attached to hydraulic jacks pulled on the spindle's front end. The loads were increased in steady increments so that the stresses and deflections could be checked against the calculated values at every stage. In this way any deviation between the figures would pick up any problems well before the spindle was placed at risk. In fact, the figures corresponded almost exactly at every stage. That said, the test itself was not without its fair share of excitement, not least when a nearby lightning strike, some three hours after the test had started, burnt out half the sensors – fortunately, after sufficient data had been gathered to show all was exactly as it should be. With the other inevitable delays, it turned into a long night and it was 4am before the test was successfully completed.

"Here we were, nine months into the project and just weeks away from the uplift, and we were deliberately over-stressing the single most critical element on the entire project. Rationally, of course, you knew all the calculations had been done and that it should all work – but you couldn't help feeling a slight risk."

Neil Thompson
Design Manager, Mace Limited

Outside on the assembly area, the A-frame legs were also nearing completion – the individual legs now welded to the top joint which was receiving its final touch of paint under a temporary shroud.

"The last section of the spindle only arrived about 10 days before we were due to do the test, which made it a great rush to get everything ready. It arrived on a Saturday afternoon, after it had been stopped twice by the police, but immediately we moved it into the small assembly hall we were using and tack welded it to the two other sections. Everything had to be surveyed and checked, so it was a long night. But it went well and on Sunday afternoon we started the pre-heating so that the we could begin welding on the Monday morning. The thickness of steel we were welding was around 140mm, much thicker than we had done at Hollandia before, so we brought in a special team of welders with the right experience. They were all people we knew well, people we trusted – we knew the welds had to be perfect the first time. They started with the inside welds which were the most difficult. Because of the pre-heating (around 150°C) it was very hot inside the spindle, so each man could only work for around 30 minutes before a change over. The welds had to be carried out continuously, working round the clock, but it was all completed on schedule, by the following Friday. Then we had to do all the testing and a final heat treatment to relieve the stresses, leaving only the Monday and Tuesday to complete all the preparations for the test on Wednesday. Then, on Thursday morning, we started the shot blasting."

Martin Broeders
Production Manager, Hollandia BV

With tolerances measured in fractions of a millimetre, the fitting of the hub over the spindle was always going to be a delicate operation, though this was made easier, to some extent, by a slight tapering of the spindle — stepping down in gradual stages from a diameter of 2100mm, at the inner bearing point, to 2040mm where the outer bearing sits. The two bearings had been sized to match. By 18 June 1999, both the hub and the spindle were complete and the bearing points properly prepared — the 'inner' point (below) finished with an area of Teflon coating so that the bearing can 'slide' over the spindle in response to minor deflections and changes in temperature — only the outer bearing is fixed. In a carefully controlled operation, the hub was first manoeuvred over the spindle (opposite). Both bearings were then preheated (left), causing them to expand slightly, so that they could be positioned over their bearing points precisely. The entire operation took a full day and was a complete success on the first attempt.

"As there was no way of guaranteeing that the two bearing points were exactly aligned — the spindle is just too big — it was decided to specify spherical bearings as these allow a slight angular misalignment between the two. With the loads we have here, this is only about half a degree, but as the bearings are so far apart this is plenty. These were our standard design in most respects. The only real change we made was to taper, very slightly, the outside of the bearings' outer rings to make it easier to remove them from the hub, should that ever be necessary. These are normally straight. This was decided quite early on, so that the hub castings could be machined at the same angle."

Gerhard Kleine
Production Manager, FAG OEM und Handel AG

With its tapered legs standing 20 metres apart at their base and measuring well over 60 metres in length, the 310-tonne A-frame was prepared for despatch to London, like the rim sections before it, fully painted and in as complete a state as possible, fitted inside and out with the appropriate fittings, ladders, intermediate work platforms, cable fixing points and, in one of the legs, a small mechanical hoist. A temporary boom held the legs apart at the correct distance at their base, while straps supported temporary walkways along the top of the legs, to allow safe access on site for as long as the A-frame was held at the horizontal. Even the backstay cables that would eventually span between the fixing brackets at the bottom of the legs and the end of the spindle were in place, tied down to the walkways, it being easier to load them now, rather than later over the river. Two floating cranes lifted the A-frame clear of the wharf *(below)*, before backing away to allow the main transport barge to be towed into place below the frame, which was then lowered on to its temporary supports *(far right)* ready for transporting to London.

Great care was taken when setting the A-frame down on to the barge, to ensure it was properly aligned and fully supported at every point — an essential precaution if all risk of distortion during the sea crossing was to be avoided.

As the A-frame legs and the hub and spindle assembly were to be lifted into place on site on consecutive days, it made sense to transport them to London on the same barge. Programme deadlines were looming and everything was now happening very quickly indeed. The loading was completed in one day, on 28 June 1999, and the barge set sail that night to arrive at its holding point in the Thames Estuary, near the Dartford Tunnel, by the following afternoon. Here, everything was washed down and checked, while ballast was removed from the barge to improve its manoeuvrability during the trip upriver to site. This took place on the very next day and, on the day after that, the A-frame was lifted into place.

L 21,50
B 5,50

Following the ground-breaking ceremony in January 1999, work on the Jubilee Gardens site and the adjoining stretch of the river began immediately, to ensure everything was ready for the arrival of the first completed sections of structural steel in June. By mid-March, the narrow site alongside County Hall had been cleared and a site office erected in the corner closest to Belvedere Road. Incorporating the Mace project office, meeting rooms, a canteen and secondary offices for the various trade con-

tractors, it quickly became the main focal point for everyone involved in the project. Meetings, formal and informal, planned and unplanned, ebbed and flowed through the confined spaces of the Mace office throughout the day and often late into the night, as members of the different contractors coordinated their operations — an especially important process on a project where site access, both on land and on the river, was so limited.

The Site

Long-time advocates for greater use of the river, Marks Barfield Architects had been determined from the very beginning that the wheel should be situated by the Thames, preferably as close to the centre of London as possible. When the options were considered, Jubilee Gardens on the South Bank stood out immediately as the obvious choice, offering the very best views over the historic heart of the city while continuing a tradition for forward-looking structures on the site that had been established during the 1951 Festival of Britain. And as more detailed research was undertaken (below) — into links to public transport and other nearby tourist destinations, as well as studies on the impact to traffic and tourist movements and so on, by Ove Arup & Partners — the more the correctness of that initial decision was reinforced. There was one restriction, however, imposed by a strategic view of St Paul's Cathedral (right), from a point on the Embankment near Westminster Pier. Strict guidelines forbid the view being blocked by any permanent structure, and the only solution was to build the wheel at the end of Jubilee Gardens nearest County Hall. As it turned out, this was the preferred site and was on land that, for the most part, belonged to the owner and landlord of County Hall, Shirayama Shokusan. The company was enthusiastic and offered its full support.

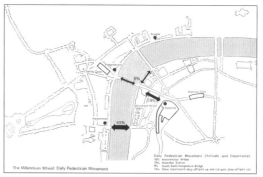

The Millennium Wheel: Daily Pedestrian Movement

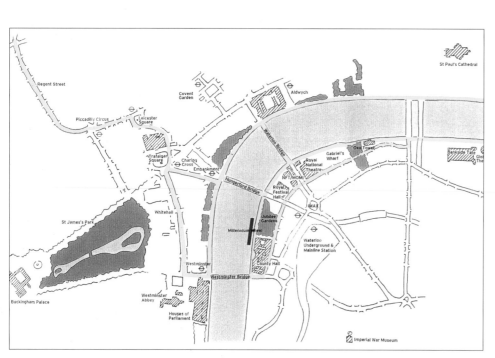

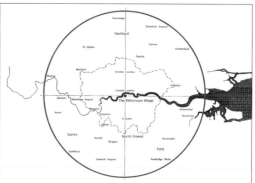

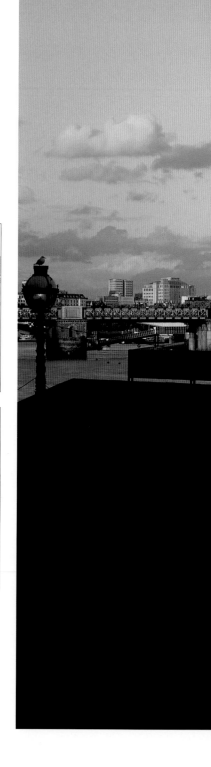

"The wheel is an imaginative way to mark the millennium, providing unparalleled, once-in-a-lifetime views of the heart of the capital. It has caught people's imaginations and created a sense of excitement about the millennium, not least among children. It will boost the economy of the South Bank and will bring visitors and new life to this important stretch of the River Thames."

John Gummer
Secretary of State for the Environment

"The more we found out about the site, the more right it seemed. The practical advantages were clear but, more importantly, there were also strong arguments that it would bring very real benefits to the area. The project could be matched to any number of government and local strategies for the regeneration of that part of London, creating local jobs and helping to draw the wealth of the north bank across the river. Not that there wasn't initial resistance. Lambeth consulted over 100 bodies of various sorts – local and statutory, as well as all of London's 33 boroughs – and over the next two years we spent a lot of time talking to them, explaining our ideas and gradually gaining their support. In the end, the response was overwhelmingly in favour, to the extent that all three chairs on the Lambeth planning committee – there being a hung council at the time – had prepared their own speeches welcoming the project. Even then we could not be sure, as the project could have been called in by the Department of the Environment, but then John Gummer gave it his support, too. Looking back now, in many ways one of the greatest successes of the project lay not so much in the design, nor in finding suitable contractors and manufacturers, but in gaining the necessary permission to build the wheel where we did."

Julia Barfield
Partner, Marks Barfield Architects

Work on the Jubilee Gardens site began in February 1999, with the general clearance of the site as a whole and the sinking of the piles that underpin the foundations, many to depths of over 33 metres. Connected by heavy underground beams to act as a single integrated unit, the foundations comprise two main elements: a compression base *(top)*, with 45 piles, to support the wheel's A-frame legs; and a tension base *(above)*, with 12 piles, that provides the fixing point for the main backstay cables. The piling was completed by mid-March, allowing excavation to begin immediately *(right)*.

For a variety of reasons — partly to do with ease of construction and partly because it was faster — it had been decided to build the wheel horizontally over the River Thames. Fortunately, this was possible as the site is located on the inside of a curve in the river at that point, and the necessary river works — including a number of temporary platforms to support the structure of the wheel as it was assembled — would not impede the main navigation channel over by the far bank. The civil engineering work itself, which included the preparation of the site, the construction of the main foundations and all the river works, both temporary and permanent, was undertaken by Tilbury Douglas Construction Ltd, a specialist in marine and river works. Appointed under a 'design and construct' contract, Tilbury Douglas selected Tony Gee & Partners to carry out the complex engineering work, while extra support in obtaining the necessary approval and permissions for the works in the river was provided by The Beckett Rankine Partnership.

Work on the river began with the sinking of piled foundations for six of the eight temporary platforms required during the assembly of the rim, and the sinking of permanent foundations alongside The Queen's Walk — with vertical piles for the boarding platform and restraint towers, and large raking piles for the main anchor blocks that now support the landside ends of two pedestrian bridges (or brows) fixed at their far end to a new ferry pier. These brows not only provide access to the pier, but also act as its main mooring points. The anchor blocks, therefore, had to be strong enough to resist the rotational forces imposed by the flow of the river on the pier itself, as well as lateral forces imposed during a possible collision. To avoid any possibility of damage, the ornate Victorian lamp-posts and the carved granite facings of the river wall were carefully dismantled and put into storage for the duration of the works *(bottom right)*, and replaced when the project was complete.

"When we were first approached in October 1998 to submit a tender for the site works, Mace had made it clear that the design was still in its early stages and that the tender was to provide no more than a benchmark price — they knew they had to appoint a contractor sooner rather than later and this was the easiest way. What we hadn't anticipated, however, was just how much the job would grow, and it wasn't really until we were formally appointed in December that the true scope of the work became apparent.

Suddenly, everything changed. There we were, just weeks before we were supposed to start on site, trying to source the extra equipment, check its availability and make sure it would be in place on time. It didn't matter that the engineering was incomplete; we just had to try and cover every contingency. I don't think any of us had much of a Christmas."

Steve Radcliffe
Divisional Director, Tilbury Douglas Construction Limited

The setting out of the piles and platforms, both temporary and permanent, was carefully monitored at every stage to ensure the structural sections they were designed to support would fit exactly. When possible, much of the surveying, painting and welding was carried out from the shoreline alongside The Queen's Walk *(below)*, which was regularly exposed at low tide.

"I came on board in December, just as the job was beginning to gear up. There was the difficulty of trying to establish the size of the piles, so we could bring in the right piling rigs and so on, but there were also a number of concerns about the state of the site. It had a long history, so we weren't really sure what we would find. We sunk some boreholes and checked the archives, but neither method is that accurate. As it turned out, we uncovered part of the foundations for the Dome of Discovery, as well as a section of cobblestone wharf and a large brick cistern that must have been part of a Victorian sawmill. We had some archeologists from the Museum of London on site for a time who unearthed all sorts of smaller remains, some going back to mediaeval times. Even the river bed was surveyed for unexploded ordnance left over from the War."

John Lovell
Contract Manager, Tilbury Douglas Construction Limited

Though less massive than the compression base, the tension base *(below)* is an equally impressive structure, requiring great accuracy during the setting out and construction of the tension chamber.

"Tilbury Douglas approached us in October to help with their tender bid, and then appointed us at the beginning of December to prepare designs for all the foundations and river works. Even at that stage it was not entirely clear what the scope of works might be. Iv-Infra were still working on the calculations for the imposed loads during the uplift of the wheel, so we had to come up with an initial design that was flexible enough to be adapted as the final conditions became known. The most difficult elements were actually the raking columns that support the A-frame. Now that the wheel is erected, they just act as radial struts and give the appearance of straightforward items, but during the uplift, particularly when taking into account the possible dynamic effects due to wind loadings, there were any number of load combinations that had to be considered. We worked closely with the architects to develop a form that was as slender as possible, but that would also accommo-date all the reinforcement we needed while leaving enough space for safe access. It was not easy to resolve but, in a way, that just makes it all the more satisfying to know we came up with a good solution."

Graham Nicholson
Partner, Tony Gee and Partners

May 1999 and the first of the permanent steel shutters for one of the raking columns is made ready for erection *(below)*. The steel shutter offers no structural advantage, but was considered the best way of combining a high-quality finish with a compact form.

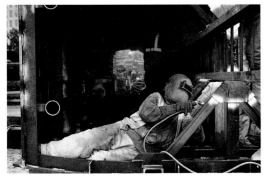

By early April, the excavation of the main foundations for both the compression and tension bases was complete and the setting out of the extensive reinforcement was well under way. Supporting the full 1800-tonne weight of the completed wheel — all transferred down to the ground through the two raking columns that support each of the A-frame legs — the compression base acts as a stiff beam, resisting the 'spreading' load imposed by the splayed legs and transferring the point loads under the columns equally among the piles. Crisscrossed by a dense web of reinforcement bars — interlocked with the pile heads below and the reinforcement for the raking columns above — the compression base was split into three equal 'cells' for the pouring of the concrete, each pour being a continuous operation taking up to five hours a time.

"Considering the scale of the project, it was a real postage stamp of a site, especially with so much large equipment about. It was a constant battle to co-ordinate all the overlapping trades — making sure everyone had access to their area of operations at the right time; that they could work safely; and that their deliveries were properly scheduled."

Andrew Elliott
Project Manager, Tilbury Douglas Construction Limited

By Tuesday 15 June, the day that the first rim section *(below)* would be lifted on to site, the pouring of the concrete for the compression base had been completed and the first of the two raking columns had been lifted into place. Erected with a flat back at this stage, a curved cowling would be welded to the permanent steel shuttering of the columns at a later date, completing their elegant form and, at the same time, concealing access ladders and supply cables. Out in the river, the support platforms for the rim sections due to arrive in the following weeks are nearing completion, while on site the tension base, with its heavily reinforced tension chamber roof *(right)*, awaits its final concrete pour.

TENSION
BASE

The arrival of Taklift 1, which was brought up river on Thursday 10 June, signalled a new stage in the development of the project, with a marked shift of emphasis from the prefabrication of the previous months, in factories all over Europe, to a new phase focussed primarily on the work in London. Over the following months, the gleaming white form of the wheel, laid out on its temporary platforms and stretching almost halfway across the river, became a major landmark in its own right, drawing tourists and Londoners alike across Westminster Bridge and along the adjacent Queen's Walk for a closer look. The decision to build the wheel horizontally could well have been stymied at the outset but, fortunately, the Port of London Authority, the statutory body with responsibility for this stretch of the river, have long been keen to promote the Thames as a working river and supported the scheme from the very beginning, offering help at every stage.

Site Assembly

One of Europe's largest seagoing floating cranes — or sheerlegs, to use its correct technical term — Taklift 1 is an impressive vessel by any standards. With the main jib raised to its full 45 metre height, it is capable of lifting over 800 tonnes, while with the standard fly-jib in place — as used on site — it can easily lift 400 tonnes at an outreach of around 20 metres, well within the requirements of this project. By far the largest vessel ever sailed this far up the River Thames, Taklift 1's journey to site was timed with the falling tide to ensure adequate clearance below the various bridges, but while sufficient downstream current was still flowing to provide the headway necessary to maintain a steady course. As a final check, small flags were raised on the outer corners of the leading frame to a position slightly higher than the vessel's highest point, with members of the crew watching to check they cleared the undersides of the lowest bridges.

"The erection process, ultimately, was driven by what size crane you could place on the site. With so much work going on in the river, land-based cranes were no help at all — so it came down to what size rig we had that could travel that far upriver, what its outreach would be, how close could it manoeuvre into the river wall and so on. Hollandia had to work with what we could supply"

Roger Wilson
Contracts Manager, Smit International (UK) Ltd

"We have done a lot of work with Hollandia over the years, so they knew our equipment very well, knew what it could do – what they didn't know about was working on the River Thames. From the loads they were talking about, we knew that Taklift 1 was the only option, which was already a big help because we had used Taklift 1 during the demolition of the old Blackfriar's Bridge some years ago – so we knew reaching that point was not a problem. We had no idea of conditions further upriver, but were able to get the infor-

mation we needed from the PLA (the Port of London Authority). *It turned out the only real problem would be the new Blackfriar's Bridge – the river is shallow there and the bridge has shallow arches. It was critical, but not impossible: we only had to pick our time carefully to match the tide. The clearances were small – only 30 to 40 centimetres above and below – but the real problem was holding everything straight. When everything was folded down, the length from the bow of the lead tug to the stern of the crane*

barge was over 170 metres, and the view from the bridge of Taklift 1 is very limited. We placed tugs either side to control the swing, but all the main decisions had to be made from the lead tug, which is why we insisted we use our own, with our own crew. Like all these operations, the real work is in the planning. If that is done well, then the work on the day should be relatively easy."

Jan Kwadijk
Project Manager, Takmarine BV

87

Working closely with the Taklift team, Hollandia knew it would be possible to sail the one-quarter rim sections through to the site as long as the moves were timed to coincide with the right tides. Even then, clearance below Southwark Bridge *(below)* was critical, with its shallow arches reducing the available headroom to as little as 40 centimetres. Realising local expertise was essential, Broedertrouw BV — the company responsible for the safe transport of the individual sections of the wheel — used their own crews only for the main sea crossing from Rotterdam, delivering the barges to a mooring just downriver from the Thames Barrier. Here, control was transferred to the shipping agents Thames & Orwell, who appointed Thames & Medway Towage to carry out all the moves further upstream, with support from General Marine and bargemaster Peter Sargent. Working closely with pilots from the Port of London Authority, this new team were able to deliver every section to site safely.

The sideways swing of the barges had to be carefully controlled not only as the rim sections passed under the lowest bridges *(above)* — to ensure there was sufficient clearance, first on one side, then the other — but also in the short distance between London and Southwark Bridges *(right)*, where the tidal flow is disturbed.

"There have been Sargents working on the river for more than 200 years. My great-great-grandfather was a Thames Waterman, operating out of Southwark, and my father and his seven brothers were all Watermen, too. But things have changed enormously in the last 30 years and now there is only my cousin and I from our generation, with only my son to take over after us. Hopefully, the tradition will continue, as you will always need people who know the river well, have the experience. The river is changing all the time and you have to learn its ways, where it can catch you out. The key thing with moving the rim sections was to use the tides. We were working with a 'dumb' barge, of course, and this would want to swing, but that can be controlled with the right number of tugs. If the tides weren't right, on the other hand, you were stuck. Hollandia had clearly done their homework and had timed all the movements to fall on the fortnightly spring tides when the river is both at its highest and its lowest. The currents are stronger, but it meant there was enough of a drop to give us the clearance we needed at every bridge. Of course, the tide charts are only a guide. The weather – if it has been raining, if it is high or low pressure and so on – all have an effect, but the PLA had posted a man at Tower Pier who was able to provide accurate readings when we needed them."

Peter Sargent
Licensed Waterman, Sargent Brothers

Timed to coincide with the high tide — so as to allow a safe draught beneath Taklift 1 as it worked close to the riverbank — the first one-quarter rim section was lifted on to the temporary supports that had previously been carefully aligned along Queen's Walk on Tuesday 15 June. Just as the rim sections had been built to very tight tolerances, the final assembly, too, required total accuracy and the whole operation was monitored continually by a surveying team from the Dutch specialist Delta Surveys. Responsible for every stage of the assembly and uplift, they ensured that the section was lowered within the set guidelines. Even though working at close to its furthest outreach for the load it was carrying (around 144 tonnes), the Taklift crew were able to position the rim section well within the tolerances allowed. Hydraulic jacks were then used for the final accurate positioning.

91

In the week following the positioning of the first rim section, two further sections were brought up river and lifted into position by the crew of Taklift 1 (below right), on to temporary platforms over the river (opposite). The rim section on the upstream side nearest Westminster Bridge was located first, followed by the section opposite The Queen's Walk and furthest out in the river. As before, each operation was closely monitored by Jan Stam and the team from Delta Surveys (below left). Final welding started as soon as the rim sections were properly aligned, followed by the final weld testing and painting. Because it had been necessary to manoeuvre Taklift 1 within the circle of the rim while these first sections were lifted, the central platform required to support the hub and spindle could only be moved into position once these operations had been completed. Piling had been discounted due to the weight the platform would have to support. Instead, at Hollandia's suggestion, it was built on a barge, which was now towed into position and deliberately sunk on to a levelled stone base prepared earlier on the river bed.

"Our biggest problem with nearly all the work on the river, but especially with the main lifts, was timing everything to coincide with the high tides. Taklift and the barges were moored well out in the main flow where we could work at any time, but once you started to move inshore time was always very limited. Before any operation, we had to be very sure we could complete the work in time — to bring the section into position, lower it exactly, disconnect all the chains and shackles, and then move out before the tide fell too far. It didn't help that the tides changed so much. Most of the big lifts, like the first rim sections and, later, the A-frame and hub and spindle, were timed to coincide with the high spring tides to give us a bigger leeway. But this only allowed four or five days — if we had missed the dates the work would have been delayed for at least two weeks. This was especially critical with the first rim section, as Taklift 1 had to move in very close to the river wall where it was particularly shallow. And, of course, we hadn't done this before, so we weren't really sure how long each part of the operation would take. As it turned out, the rim sections were well balanced and we had good supports in place, so the work went very smoothly."

Jan Stam
Site Operations Manager, Hollandia BV

Connected at its base by pin-joint connections with tolerances measured in a few millimetres, the final surveying and positioning of both the cast steel feet atop the raking columns and the central platform out in the river had to be carried out with extreme accuracy to ensure the massive A-frame legs would fit exactly. At this stage, the raking columns were still wrapped in scaffolding, partly to allow access and partly because their permanent steel shuttering was still awaiting shot blasting and its final coats of paint. On the day of the lift, Taklift 1 made light work of manoeuvring the A-frame into position, and all the careful surveying paid off when the pin-joint 'flanges' at the foot of each leg dropped snugly into place. Temporary pins allowing a degree of movement were inserted *(bottom left)* to hold the legs in place during the fitting of the hub and spindle. Only when that operation was complete, and the whole structure properly aligned, were these removed and the permanent stainless-steel pins *(bottom right)* fitted in their place.

The scale of this job is such that you have to think about things you wouldn't normally worry about. The A-frame legs, for example, are so long that even a modest load will cause them to contract a little under compression. It is not difficult to work out — it's around 20mm now the wheel is complete — but it had to be taken into account."

Allan Mann
Technical Director, Babtie Group

"We couldn't be entirely sure just how much the central support and its barge would settle after the loads had been put in place. The combined weight of the hub and spindle assembly and the A-frame legs was well in excess of 600 tonnes so we had to assume the barge would sink into the river bed by a significant amount, even if the stone base we had laid would help spread the load. In fact, it ended up settling around 600mm, but fortunately this was within the scope of the hydraulic jacks installed by Hollandia to correct the difference."

John Lovell
Contract Manager, Tilbury Douglas Construction Limited

By the time the A-frame legs were lifted into position, all foundation work on the tension and compression bases had been completed. Out on the river, the base frame of the boarding platform, painted white, was already in position alongside The Queen's Walk, and work was continuing on the raking foundation piles that now support the landside end of the brows leading to the new ferry pier. Just before the lift commenced, members of the assembly team — from Hollandia, the Mace site office and the Taklift crew — lined up in front of the hub and spindle for a group portrait (bottom right), mirroring a photograph taken more than a century before that had recorded the same moment during the erection of Walter Bassett's Great Wheel in Blackpool (top right).

Carefully timed so that the final positioning coincided with the mid-afternoon high tide, the hub and spindle was lifted clear of its transport barge *(right)* well before noon, allowing plenty of time for Taklift 1 to manoeuvre itself carefully alongside the central support platform. To simplify the operation, the hub and spindle was lowered on to Teflon coated plates some 500mm forward of the A-frame legs and positioned accurately later using hydraulic jacks.

At 335 tonnes — only slightly less than the 'dry' weight of one of British Airways' Boeing 747-400s — the completed hub and spindle assembly was the single heaviest component shipped to London and lifted into place on site. It arrived with its own support frame fitted to one end, which not only formed the base on which the whole assembly would stand during the fitting of the spoke cables, but also helped to hold the hub in place during the lifting operation, and thereby avoid excessive lateral loads on the bearings. As was to happen on so many key moments during the wheel's assembly and uplift, the day of the lift was blessed with near perfect weather: with clear skies, warm sunshine and almost no wind.

"The decision to build the wheel horizontally was presented at our first big meeting with the client team in September 1998. We knew enough about the fabrication by that time to know that trying to complete the wheel by December 1999 was just not possible if we had to build vertically — you had to raise the A-frame and the hub and spindle before any real work could begin on the rim, and there just wasn't time. Apart from all the practical advantages, building the wheel horizontally allowed work to be carried out in parallel, which saved us at least a month — if not two. With our experience of moving other large structures, we knew that the weights we were dealing with would not be a problem in themselves, but going in that direction would clearly have a major effect on the way we designed the structure. We knew building horizontally was a good solution, but it was only possible if we could use the river — and until we knew if this would be allowed, we had no idea which way the design should proceed. We really needed an answer as quickly as possible."

Peter Koorevaar
Technical Director, Hollandia BV

The lifting of the last of the major structural elements was timed to coincide with the high spring tides at the beginning of July, a necessary precaution as Taklift 1 had to work close in to the shore. The rhythm of the tides allowed a five day period of opportunity. The A-frame legs were lifted first, on Thursday 1 July, followed the next day by the hub and spindle. A temporary yellow access bridge, spanning between the top of the spindle and the rim, was positioned over the weekend and one final river platform — to support the centre of the last rim section — sunk into place on a barge. With the Monday given over to surveying and final adjustments, the wheel was finally closed with the positioning of the final rim section (*main spread*) on Tuesday 6 July.

Following the lifting of the last of the main structural sections, a full overall survey of the wheel was carried out and the different elements positioned precisely, allowing the final welding to begin and the permanent pins to be inserted in their respective pin-joint connections. With everything in place, work could then commence on the next major operation: the fixing of the spoke and rotation cables. Unsurprisingly, for a structure where many construction procedures had never been attempted before, the introduction of each new operation was greeted with varying degrees of uncertainty. It often took a few days to refine working procedures that had looked straightforward on paper, and so it was with the fitting of the spoke cables. During the weekend of the final lifts, Hollandia had positioned a temporary access bridge spanning between the top of the spindle and the rim *(left)*, to help with the lifting of the spoke cables. Unfortunately, the system proved unwieldy in practice and the first cables were not fitted *(below)* until late July.

"I can still remember clearly that first big meeting with Hollandia in Rotterdam, when they said they would like to propose a new erection method. The whole team was there, all the usual suspects, looking at these drawings that showed the wheel laid out across the river. None of us said very much, other than to say we weren't really sure about it. We moved on to other issues, but when we left there was a definite feeling that this might be the end of the project. But then, on the the train back to Amsterdam, we were sitting there huddled across the centre aisle chatting about it, and the more we talked, the more we thought it might just be possible — maybe we could close the river when we needed. By the time we reached Amsterdam the mood had changed completely, from a real low to a real high. It was certainly an interesting trip and, somehow, it seemed to set the pattern for the year to come."

Andrew Potter
Project Coordinator, The Tussaud's Group

All the cables for the project were manufactured by the Italian company Tensoteci s.r.l. at their Redaelli factory near Milan. Varying in diameter from 60mm for the rotation cables to 110mm for the main backstays, each cable is made up of individual strands – circular at the centre and 'S-shaped' to the outside – spun in layers around a central core in a carefully controlled operation that ensures each layer is tightly bound to its neighbour. Shown opposite, at true size, are sections through two of the cables used on the wheel: a 70mm spoke cable, made up of 121 strands in six layers, and a 110mm backstay, with 303 strands and 10 layers.

"Cables have been made in the same way for many, many years, so much of the engineering work is now well understood – the calculations can be very lengthy but performance can normally be predicted with great accuracy. The real challenge is in understanding the dynamic effects due to wind loadings. There are two main conditions to consider: the effects of vortex shedding, where the wind causes resonant vibrations to build up in the cable – this is normally solved using standard Stockbridge dampers – and an effect known as wind and rain 'galloping', which occurs when rain-water collects on the cable and forms a rivulet down one side. This moves as the cable moves and amplifies the vibration. Usually the rivulets are broken up by winding an extra raised wire along the length of the finished cable, but we were worried this might be damaged due to the constant flexing on the wheel. As it happened, there was an alternative solution we had been thinking about for some time, which involved spinning thinner strands into the final layer of the cable to form spiralling grooves. We knew it was a good solution, but it was a bit of a gamble to use them on the wheel as such a design had never been used before in practice. As it turns out, they are working very well."

Massimo Marini
Technical Director, Tensoteci s.r.l.

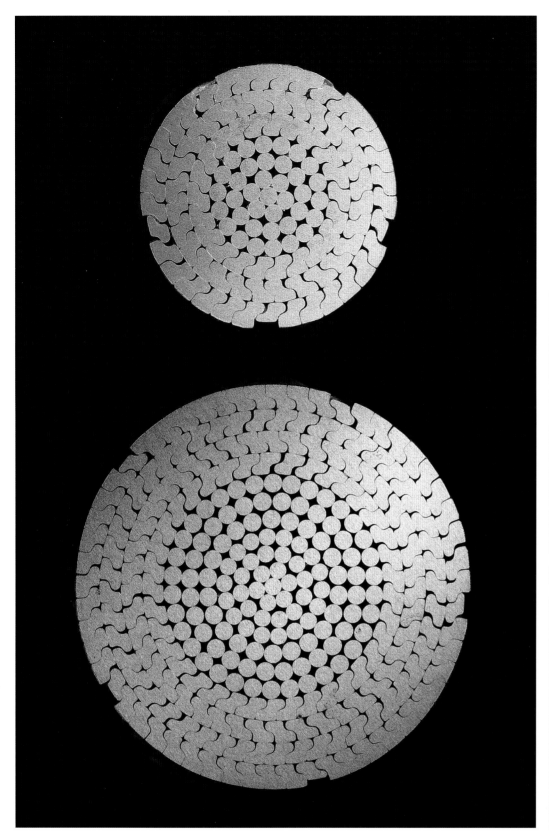

Samples of each of the types of cable used on the wheel were tested in Germany at the DMT Institute *(bottom left)* – an authorised body for the non-destructive and destructive testing of all forms of ropes, wires and cables – on purpose-built machines capable of applying loads of 2000 tonnes or more: certainly far greater than the loads ever likely to be applied to any of the London Eye's cables.

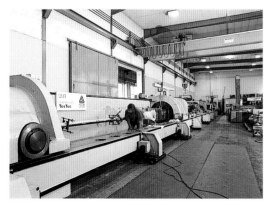

"Tensoteci appointed us to carry out the usual range of destructive and non-destructive tests on a sample of each of the cables used on the wheel, not so much to provide data to help in the design process – modern cables are designed very accurately nowadays – but more as a way of providing an independent verification of their calculations. The cables were supplied in lengths of around five metres, complete with their standard fittings, which allowed us to measure the amount of creep in the fittings as well as determine each cable's modulus of elasticity. We followed an established procedure, increasing the load in increments up to the usual working load for each cable – normally around 40 per cent of its breaking point. When we were happy that this had been accurately recorded, we then moved on to the final stage, loading the cable until it broke."

Friedrich Dürrer
Head of Test Bay, DMT Gesellschaft für Forschung und Prüfung

"The general design conditions for the different cables were specified by Hollandia, but the detail design, the choice of dimensions, the fatigue calculations, the dynamic analysis and so on were all carried out by Tensoteci. If there were any concerns about the use of cables, they lay in their reaction to dynamic wind loads: whether vibrations in the cables would cause discomfort for the passengers in the capsules. In the same way that we verified the design of the main structure, we therefore asked the Engineering Office of the City of Rotterdam — who have good experience in solving the problems of cable vibrations — to review Tensoteci's proposals and, where necessary, make their own recommendations. The final design seems to be very successful and, if anything, vibrations in the cables are even lower than expected."

Chiel Smits
Project Manager, Hollandia/Mercon Steel
Structures BV

The basic techniques employed in the making of cables — winding layers of individual strands around a central core, the first layer spiralling one way and the second the other — have remained essentially unchanged for more than 100 years. Certainly, many of Tensoteci-Redaelli's winding machines, including the one shown here spinning the main 110mm backstays, have been in use for 40 years or more. This is not to say advances have not been made, but in recent years these have been mainly in the better understanding of the engineering principles and in subtle refinements, notably the introduction of the 'S-shaped' strands that make up a cable's outer layers. Tensoteci are, therefore, rightly proud of their own innovation: the introduction of narrower strands to create grooves in the cable's outer surface. Manufactured in continuous lengths, several hundred metres long, the completed cables were first transferred to a purpose-designed 'runway' where they were stretched to their full working load and cut to length — all this after several stretches at higher loads had been applied to ensure all non-elastic deformations had been removed. In a final operation *(see previous page)*, their final connections were then drawn over the cables and fixed at each end.

"The strength of the cables is not in question. Predicting their life, however, is a little more difficult. With the main backstay cables, where the load conditions are more or less constant, we can safely predict a life of 50 years with good maintenance. For the spoke and rotation cables, however, we are predicting around 20 to 25 years – though this will be monitored closely."

Massimo Marini
Technical Director, Tensoteci s.r.l.

"Of all the lifts we had to make during the assembly of the wheel, those for the cables were actually the most problematic. With all the big sections, it was mainly a matter of connecting them to the crane and balancing them correctly, after which the crane did all the work. With the cables, all the lifting had to be done by hand, using secondary cables, hand winches and chain-blocks. Finding good fixing points for the winches took time and the scaffolding was never where you wanted it. It did not help, either, that the temporary 'flyover' bridge we had installed did not work as well as we had hoped – again, it relied on too many winches, which just took too long to rig and adjust. It did the job, but it was very slow – so we had to work out another method, here on site, that could work in parallel. That is when we decided to use an extra barge, specially adapted for the job, on to which we could load the cables and then float them under the wheel to where they were required. This worked well but now we were reliant on the tides, so it could not be used all the time. It was very frustrating and the entire operation ended up taking far longer than we had planned."

Jan Stam
Site Operations Manager, Hollandia BV

By the beginning of August the fitting of the spoke cables was progressing, though not as smoothly as had been hoped. The problems involved in lifting the cables horizontally while avoiding excessive bending — which could cause irreparable damage — was proving more difficult to solve than first imagined, and the temporary 'fly-over' bridge (right) that was meant to make the job easier was too slow. To speed up the operation, Hollandia decided to introduce an alternative fixing method, lifting the cables directly from a barge floated into position beneath the wheel (bottom left). Weighing around 3.4 tonnes each, complete with all their fittings, the cables were first pin-jointed at the hub, then the far end pulled into its appropriate fixing bracket on the rim where it was secured with a threaded collar (below).

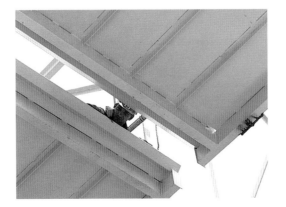

"I am not sure if I should tell you this, but we did lose one of the cables when the lorry transporting it from Italy collided with a bridge. Fortunately, we had already decided to make a spare cable of each type — three, in fact, for the spoke cables — so there was no delay and its replacement, the new spare, was completed before the wheel opened to the public."

Adry Zondervan
Quality Assurance Manager, Hollandia BV

The fitting of the spoke and rotation cables was eventually completed towards the end of August, some weeks later than planned but otherwise with complete success. Measuring some 53 metres in length, there are 64 spoke cables in all, 32 spanning radially from each of the hub castings to alternate, equally spaced fixing brackets along the rim's inner chord. As each cable was fitted, a modest tension was introduced to hold the cable horizontal. Later, once all the cables were in place, this was then increased — in a carefully balanced operation which ensured the central hub and spindle was not pulled out of place — to the cables full working load of around 75 tonnes. When the wheel is vertical the weight of the rim is naturally carried by the lower cables alone, reducing the tension in the upper cables. In a high wind, therefore, there is an increased risk that the cables on the leeward side of the wheel might go slack, allowing increased movement around the connections with the potential for irreparable damage. The pretensioning precludes such a situation occurring. A further 16 rotation cables, each 60mm in diameter, were also positioned at this stage. Running at an angle to the spoke cables to fixing positions on the outside of the rim, these ensure there is no lag between the turning of the rim and the turning of the hub.

"At the time the hub castings were designed, we were still assuming the rim would be supported by just 32 cables, working in pairs. This was eventually changed to 64 individual cables, which doubled the number of pin-joint connections at the hub. Fortunately, the castings had flanges large enough to accommodate these without changing the design. Sometimes you need a little luck."

Professor Jacques Berenbak
Senior Design Engineer, Hollandia BV

"The amount of available space in the fixing brackets on the rim was quite limited, so the tolerances set for the lengths of both the spoke and rotation cables was really very small – no more than plus or minus 50mm. As the expected creep in the cables during their lifetime will probably be around 20mm, it was best to work to the low side, and considerable care had to be taken when stretching the cables at the factory, and then cutting them to their correct working length, to make sure the they were fully bedded in and variables such as the temperature had been taken into account. We had decided at the outset to fit permanent strain gauges within the pin-joint connections at the hub of every spoke and rotation cable, in part to monitor the load changes as the wheel rotates – which will help confirm our fatigue calculations – but also, and more importantly, to check that even under the worst conditions none of the cables ever goes slack. As it turned out, the sensors had an added benefit, as we were now able to monitor the loads on the cables not only during the pre-tensioning – which allowed us to ensure the overall loads remained balanced at all times – but also later, during the uplift."

Massimo Marini
Technical Director, Tensoteci s.r.l.

With a total weight of around 1200 tonnes at the time of the lift, British Airways London Eye is relatively light when compared with lifts achieved elsewhere, but it is particularly unwieldy. Its large diameter made it impossible to lift vertically from the centre, while its height when upright ruled out any chance of lifting at the rim. By chance, in the months preceding their involvement in the wheel, Hollandia had been commissioned by Mammoet Engineering, a specialist in heavy lifting operations, to make special sections for a very large mobile 'crane' — actually a sliding gantry — that was more than capable of handling the loads involved. Working together, Hollandia and Mammoet devised a lifting strategy that involved raising the boom of the sliding gantry upright over the same raking columns that support the permanent A-frame legs of the wheel to form a second temporary A-frame, and then hauling the wheel upright with cables fixed securely to the massive pin-joint that connects the legs to the spindle.

The first sections of the Mammoet sliding gantry, the yellow box-trusses that form the main boom, started arriving on site in early August, just five short weeks before the proposed date for the uplift of the wheel during the second week of September. A true mobile crane, despite its massive size, the unique advantage of the Mammoet system is that it has been designed to be broken down and transported in pieces no larger, or heavier, than a standard container. Arriving in sections over several days *(below and far right)*, the main boom began to take shape on supports set out across nearly the full length of the site *(right)*. Capable of lifting loads in excess of 3000 tonnes in its full glory, the uplift was well within the system's capabilities, even though pulling the wheel upright at an angle — effectively rotating the entire structure around the pin-joints at the foot of its A-frame legs — meant that the full load on the cables as the wheel came clear of its supports would exceed 1900 tonnes.

Assembling the boom on so cramped a site proved to be a major challenge. With other trades requiring access at the same time, the number of cranes and the crane-time allotted were severely limited, and an operation that would normally have taken the Mammoet team a matter of days ended up taking closer to three weeks.

While the Mammoet A-frame began to take shape on the site above, preparation work was proceeding rapidly in the tension chamber below, commencing with the fitting of the steel anchor blocks *(above)*, which had been designed to support both the permanent backstays of the wheel itself and the guide-plates that support the Mammoet A-frame.

With the anchor blocks in place — fixed precisely to the ceiling of the tension chamber — the jacking systems required to draw down and support the main backstays were lifted and bolted into place (below). At the same time, atop the raking columns (bottom left), high-strength grout was injected beneath the cast-steel feet that support the permanent A-frame to form a solid base capable of resisting the extremely high forces imposed during the uplift.

"We were approached by Hollandia in February 1999, initially to do no more than supply and pre-tension the auxiliary cables, required during the uplift to provide extra support to the rim. Mammoet had no time to do the work themselves but, as we supply their strand-jacks, they suggested Hollandia speak to us directly. The lift procedures were still being finalised at that stage and it soon became clear that there were, in fact, several other areas where our experience with hydraulic jacks and control systems would come in useful — particularly as for most of the work we would be working alongside Mammoet, integrating our two control systems. Within a few weeks, the scope of the work doubled to take in not just the auxiliary cables, but also a complete stability system for the rim and a secondary jacking system to draw the main backstays down into the tension base as the uplift proceeded. We ended up installing two sets of jacks in the tension base: a permanent set mounted directly on the steel anchor block to hold and adjust the backstays once they were supporting the full weight of the wheel; and a secondary lighter set, mounted behind the main jacks, to actually draw the cables down during the lift."

Tjerko Jurgens
Managing Director, Hydrospex Cylap BV

Due to the extremely large loads and dimensions involved, the uplift itself was carried out using four hydraulic strand-jacks *(below)* fixed to the top of the Mammoet A-frame. Capable of lifting 600 tonnes each — the load being spread across 36 individual strands, each 140 metres long — the jacks work in tandem and automatically balance the overall load between them as the strands are drawn in. The A-frame, the jacks and the purpose-built 'head' that guides the free ends of strands out of the way, after they have been drawn in, was assembled as a complete unit while still lying horizontally across the site — the strands themselves being fitted after the base of the Mammoet A-frame had been raised up and placed on special castings fixed to the top of the raking columns *(opposite)*. Positioned over the pin-joint connections that support the bottom of each of the wheel's A-frame legs, these temporary castings — made specifically for this one operation by Skoda — held the base of the Mammoet A-frame in place as it was swung upright into its final working position.

"As the design engineer in charge of erection procedures, it was my role to co-ordinate between Hollandia, Mammoet and Hydrospex and decide which company would be responsible for each part of the operation. It all came about, of course, because we were actually making special parts for the second of Mammoet's large sliding gantries, though I wouldn't say that was the reason we thought of building the wheel horizontally — there were a number of other benefits with that approach. Once that decision was made, however, it was clear that the Mammoet system, and in particular the boom we were making, might offer a good solution. We approached Mammoet to check if the boom might be available in June and they replied that it was — as long as we completed the work on time. From that point on, we started to work out how the job could be done."

Nardo Hoogendijk
Design Engineer, Hollandia BV

"One of the biggest problems with the uplift was that the highest forces occurred at the first moment of the lift. The critical point was at the top of the 11-metre high raking columns, where the horizontal forces imposed through the permanent A-frame and the vertical forces from the Mammoet A-frame above were transferred down into the columns themselves. We had considered supporting the Mammoet A-frame at ground level, but it was actually more efficient to stand this on the columns as well, so that its vertical load combined with the horizontal forces imposed during the uplift to create a resultant force acting at 65 degrees, which was much easier to sustain."

Nardo Hoogendijk
Design Engineer, Hollandia BV

Before the Mammoet A-frame was lifted upright, the pin-jointed guide-line plates connecting the top of the frame to the tension base — and thereby preventing the entire structure falling forward into the river — were folded and positioned underneath the strand-jacks (opposite).

Following a site visit by Bob Ayling (bottom left), two very large mobile cranes were brought on to the site and the head of the A-frame raised sufficiently (below right) to allow the strand-jacks to take up the slack and complete the lift themselves.

"Mammoet introduced the first very large sliding gantry some two years ago, so we have built up a good experience of its use in normal operations — of its standard assembly procedures and so on. In fact, one of the attractions of this project was that it was so different; it was interesting to explore just what the standard equipment was capable of. As it turned out, the uplift itself was the least of our worries: the maximum imposed load on the system at the moment the rim lifted clear of its supports was no more than around 1900 tonnes, well below the capacity of our strand-jacks, while the boom itself can accommodate compression loads of up to 4500 tonnes. Our real problem was assembling the system on such a restricted site. Normally we have the site to ourselves and we would expect to assemble and raise the boom in a matter of days. Here, we had to plan our operations very carefully and co-ordinate with a number of other trades; the times when we had access turned out to be quite limited; and, of course, some of the work had to be carried out over the river. In the end, it took us three weeks just to complete the assembly. Then, there were a few problems raising the boom to vertical — again because of the small site. After that, at least from our point of view, the uplift itself was quite straightforward."

Ronald Hoefmans
Project Engineer, Mammoet Engineering &
Innovation BV

Dawn on Friday 10 September and the golden glow of the sunrise is reflected in the river. All the necessary preparations for the uplift were now complete: the final checks on the structure and lifting systems had been made and a beautifully sunny day with only mild winds had been forecast. Within the hour, the uplift was under way.

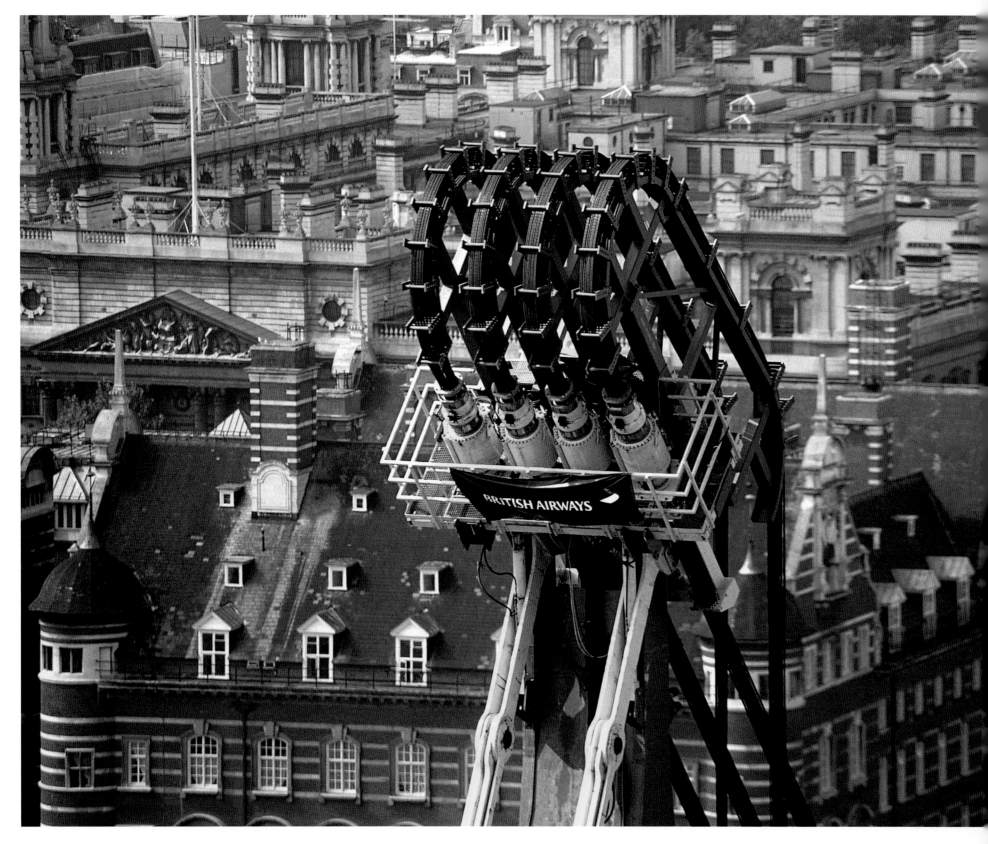

"The run-up to that first lift day was incredibly stressful. For about four weeks we had gone out of our way to court the press, both national and international, arranging visits to site and so on, getting them interested in covering 'the big day'. In the end, we had accreditation for around 200 organisations, with some 20 television stations broadcasting live for at least part of the day. Quite frankly, it was quite a surprise, as interest in the wheel before then had been quite low-key. It certainly raised the tension, and I spent quite a lot of time in the site office during that last week before the lift, trying to impress on people how important it was that everything went smoothly on the day. As I remember it, we finally received word that the lift was 'on' on the Thursday morning, just a day before the lift, although that particular date had been 'pencilled in' for some time. I was already camping out at the Marriott Hotel (in County Hall) by then, as we were working very long hours indeed during the last few days – making all the final arrangements, making sure everybody knew it was happening and so on. My one wish for the lift itself was that it would all pass off smoothly, so that we could just let everyone enjoy the show and have a nice easy day. Of, course, it didn't quite work out like that."

Jamie Bowden
Official Spokesman, British Airways London Eye

A methodical sequence of operations had been established for the up-lift — the load in the Mammoet strand-jacks *(left)* being increased in regular increments, with pauses at each stage to analyse the measurements being received from the sensors positioned all over the structure. All seemed to be going well, with the measurements corresponding almost exactly with the computer predictions. As one final step, it had been decided to deliberately overload the system by increasing the load in the strands to 110 per cent of that required to lift the wheel, actually raising the hub and spindle *(below)* while the rim was still held down on the support platforms. This increased the load on the auxiliary support cables spanning between the top of the spindle and the rim, and it was at this moment — with the media looking on — that one of them suddenly broke lose, the jolt causing the cables on either side to follow suit *(bottom left)*. The lift was stopped immediately.

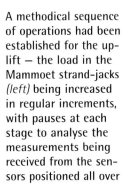

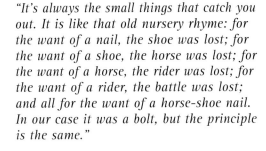

"It's always the small things that catch you out. It is like that old nursery rhyme: for the want of a nail, the shoe was lost; for the want of a shoe, the horse was lost; for the want of a horse, the rider was lost; for the want of a rider, the battle was lost; and all for the want of a horse-shoe nail. In our case it was a bolt, but the principle is the same."

Allan Mann
Technical Director, Babtie Allott & Lomax

"Engineering is not just about the calculations; there is a practical element too. Good engineering, for example, should be insensitive to defects in workmanship — especially on site where one has to accept that particular operations cannot always be carried out exactly as planned. In that sense, of course, the failure during the first uplift was the result of poor engineering, but it would be wrong to say that it was any one person's fault. Mace and ourselves had recognised from the outset that with so much pressure on the project, mistakes were bound to sneak in. There was no point in trying to apportion blame — it was better by far that people be allowed to put all their energies into finding a solution. The curious thing was that in making it clear no one was to blame, everyone felt responsible. And so it was with the test. In a matter of days we were able to identify exactly what had happened and Hollandia had designed a possible solution — in fact, they had already machined an alternative fitting."

Allan Mann
Technical Director, Babtie Group

Working through the weekend, Hydrospex Cylap, the specialist contractor responsible for all the hydraulic lifting and tensioning equipment, built a test-rig in an attempt to duplicate the failure. It took a number of attempts, and then only by exaggerating the conditions. It had been noted that some of the brackets supporting the cable connections in question had not pulled straight as they came under tension, and it transpired that, under particular circumstances, the eccentric loading this misalignment imposed could force the connection's fixing collar out of line sufficiently to lose purchase on the bracket and allow the cable's fixing block to break lose. It was now a relatively simple task to redesign the collar and bracket to ensure the same sequence of events could not happen again.

There can be no denying that the failure to achieve uplift at the first attempt came as a significant shock to all concerned. Paradoxically, the test had served its purpose and safely detected an unexpected weakness in the system. And, if nothing else, it had confirmed the stability of the main structure which had been entirely unaffected. The only embarrassment was that the one major set-back the project was to endure, in what had otherwise been a year of success, should have occurred in the full glare of media publicity. Sensibly, instead of rushing ahead, the decision was now made to step back and carefully reassess the situation and to see what else might be improved. The uplift was postponed a full month to allow everything to be checked thoroughly and the auxiliary support cables and their connections to be replaced *(right)*. At the same time, the procedures and schedules were reconfigured to allow other work to continue smoothly.

By early October, all the temporary cables had been replaced and the new fixing points at the rim checked. Other procedures had been improved and, at a final team meeting attended by all those directly involved in the lift, it was confirmed that the revised uplift would now take place over the weekend of 9 and 10 October. The lift itself was controlled via computers from a site office shared by Mammoet and Hydrospex *(below right)*, with all the relevant information relayed to a 'repeater' room in the Mace site office *(below left)* where it was monitored by the project's senior design engineers. By the evening of Friday 8 October, all the testing had been successfully completed, including a new overload test on the temporary cables. Everything was in place and the following morning, at around 9am, the wheel inched its way clear of the support platforms. As with all the pretesting, the lift itself was a gradual process during the early stages, with frequent stops to study the information from the monitoring equipment, adjust the stability cables fixed to the underside of the rim *(right)* and carry out regular surveys of the structure to ensure it was performing exactly as predicted.

"The most telling moment in the lead-up to the second uplift was coming across a relaxed Maarten Jongejan and his Dutch compatriots on the Friday afternoon, playing football in Jubilee Gardens against a team from Mace. I realised then that there was no way we were going to be in for the kind of day we had endured the month before."

Jamie Bowden
Official Spokesman, British Airways London Eye

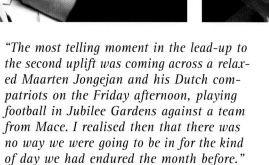

"The success of the uplift was based purely upon teamwork — and what a team it was. During that last week, Hollandia, Babtie Allott & Lomax, Iv-Infra, Tony Gee & Partners, Hydrospex, Mammoet and, of course, our own team from Mace were all on site, working together to close out the last of the outstanding engineering issues, all working very long hours with lots of inspections and lots of open discussion. And, in every instance, it was the very best people, the most senior designers and engineers. But there were no airs or graces. If an inspection had to be at four in the morning, they were there — striving to solve problems and guarantee the lift of the wheel; and not just looking to their own responsibilities, but helping each other out, always looking to achieve the common objective. And it was the same with the contractors. If further designs were needed, proposals were submitted, discussed, amended, agreed, installed and signed off — often all on the same day. Looking back, it is incredible just how much work was cleared in those last three or four days — and how much real confidence was gained. It was quite magical. Truly inspirational."

Neil Thompson
Design Manager, Mace Limited

The raising of British Airways London Eye to its upright – though not, at this stage, vertical – position took place over two days. During the first phase on Saturday, the structure was raised to 35 degrees *(below right)*, at which point it was 'locked off' for the night to allow the secondary stability cables, running between the underside of the rim and the temporary support platforms in the river, to be reconfigured. The uplift recommenced early on Sunday morning *(right)*, progressing steadily through the day and reaching its final position of 65 degrees, the permanent angle for the main A-frame legs, at around 5pm *(far right)*. Even at this incomplete stage, the top edge of the rim was some 110 metres above the river and was clearly visible from all over London. The next stage was to swing the rim itself out to the vertical, but this could only be achieved after a series of intermediate operations had been completed. These took several days and the wheel only reached its fully upright position a week later.

After the disappointment of the first uplift, publicity for the second attempt had been a little more guarded and press interest, certainly on the Saturday, was noticeably more low key. As the lift progressed, however, and reports started to appear on the news bulletins, more and more people began to arrive on the Embankment, Westminster Bridge and in Jubilee gardens to enjoy the show. It was, perhaps, at its most impressive early on Sunday morning, suspended out over the river at a seemingly impossible angle — as seen here (left) from the exposed shoreline below the Hungerford railway bridge. Note the all but completed restraint towers moored out in the river on barges alongside Taklift 3, waiting to be positioned under the rim in the following days while the wheel was held back at an angle of 65 degrees.

Measuring 110mm in diameter and weighing around 19 tonnes each, there are four backstays in total, all of which had been pin-jointed at the top of the A-frame before the uplift began. Drawn down into the tension chamber in part during the uplift, the operation was completed on Monday morning, taking great care not to damage the threaded end connectors as they were lowered through the steel anchor block (right). Once in place, the permanent hydraulic jacks set on to the back of the anchor block (above) gradually increased the tension in the cables to their full working load of around 550 tonnes, at which point the cables were 'locked off' by tightening threaded collars previously fitted over their end connectors. With a breaking load for each cable in excess of 1100 tonnes, the entire wheel can be safely supported by two cables alone.

Following a full survey to confirm the wheel's A-frame legs were held securely at the correct angle *(below)*, the support of the entire structure was transferred from the Mammoet strand-jacks to the permanent backstay cables.

"Hydraudyne was responsible for the installation of the main drive systems, the guiding roller 'sets' that provide lateral stability, and all the necessary control and monitoring systems. As conceived by Croese, the drive system is actually made up of four separate drive units, two on each side of the rim, each with four drive wheels operating in pairs that grip continuous 'running beams' fixed along each side of the rim's outer frame. In normal operation, all 16 wheels will run in unison, but the system has been designed with sufficient capacity to allow individual pairs of wheels to be retracted, should a problem occur, with no effect on the running speed. The wheel can be run normally with only 12 wheels in operation and can be safely evacuated with as few as eight, though turning at a slightly lower speed. The running beam has a high-grip coating and each pair of wheels is fitted with sensors that increase the drive pressure automatically should any slippage be detected. It is a fairly standard system and is considered very reliable. Even so, we have built in a high level of redundancy that should guarantee near-permanent operation. There are two separate hydraulic supply lines, for example, and each drive unit can be isolated and run independently – off a separate compressor if necessary. And as a final back-up, should all hydraulic pressure be lost, electric motors and mechanical brakes have been installed within the hub of each wheel. By Hydraudyne standards, this really is quite a small system, and the

Though not entirely stable in this position, the rim was held back against the A-frame legs in the week following the uplift to allow unimpeded access to the boarding platform and its immediate surroundings. This greatly simplified the lifting into place of the last of the major elements: the prefabricated control and electrical rooms on the boarding platform and, more importantly, the two main restraint towers, positioned at either end of the platform. One of these accommodates the main drive system, while both provide lateral support against wind loading and help guide the capsules past the boarding and alighting areas. Each tower takes the form of a pair of working platforms, between which the rim passes, raised high above the river. The towers were lifted into place on successive days by Taklift 3 which, like its larger brother Taklift 1, had been sailed over from The Netherlands specially for this project. The towers were positioned as finished components, with both of their working platforms — which because of their curved form were known as bananas — and all of their systems in place.

real complexity of the job — apart from controlling the logistics of such a short design and installation programme — was actually the design of the control and monitoring systems. Safety for the passengers and the operating staff was paramount, so it was essential that the systems were as simple to use and understand as possible, while foolproof protocols were also built in to ensure dangerous operations were not possible. It wasn't exactly helpful when the uplift was delayed and we had to complete the work out in the river, where access and power were limited. Still, we were determined that we would not be the ones to let the project down. If anything, it became a matter of honour that we complete the work on time, so it was very pleasing to have the systems switched on just days after the swing to vertical had been completed."

Bart Wiegmans
Project Manager, Hydraudyne Systems & Engineering BV

Capable of lifting loads in excess of 200 tonnes at an outreach of 20 metres, Taklift 3 — though significantly smaller than Taklift 1 — remained an impressive sight and was guaranteed to draw large crowds of spectators, particularly when manoeuvring the restraint towers close in to the river wall *(below left)*. As with all the structural steelwork, careful setting out during fabrication and assembly ensured the restraint towers, complete with their pin-jointed diagonal bracing, fitted on to their respective bases perfectly at first attempt *(right)*.

"For the past 10 days a festival has been going on on the banks of the Thames. People are stopping, gawping, pointing, snapping their compact cameras and setting up their long-lensed ones on tripods, swapping newly minted urban myths and, above all, grinning in spontaneous expressions of pleasure. After decades of architects and politicians blathering about bringing life back to the river, it has actually happened. The thing they are grinning at is the raising of 'The Wheel' . . . In this year of grand circular phenomena, the raising gives the most simple and direct satisfaction, less anticlimactic than the eclipse, more spontaneous than the Dome. It is a true people's monument, to use an idea that needs rescuing from the clutches of Tony Blair's wordsmiths. The wheel is a big toy, something graspable and instantly comprehensible, whose lifting into place is an action of childlike simplicity. The addictive thrills of building, of flirting with disaster before eventual triumph, which are usually known only to the consenting adults of the construction industry and hidden behind hoardings, have here been shared by everyone. The fact that the first attempt to raise it failed and that it has been teetering at a frightening angle for this past week have all been part of the drama."

Rowan Moore
The Evening Standard, 18 October 1999

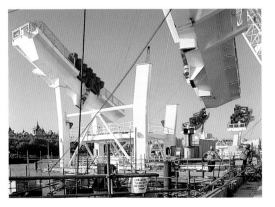

As soon as the restraint towers had been bolted securely to their base frames, the landside 'bananas' were removed to provide clearance for the rim as it was swung into the vertical position.

"The most demanding job for Taklift 3 was replacing one of the working platforms. This had been removed easily enough when its restraint tower was first positioned (left), but once the rim was vertical it could only be lifted back by slipping Taklift 3's flying jib between the spoke cables. A couple of these were moved to one side, but the manoeuvre still demanded absolute precision."

Jan Kwadijk
Project Manager, Takmarine BV

143

By the evening of Wednesday 13 October *(left)*, just three days after the first phase of the uplift had been completed, British Airways London Eye was supported on its own backstay cables and the two restraint towers had been lowered into place. Its work complete, the Mammoet A-frame, which had stood guard over the site for the past five weeks, was being lowered using the same cables that had previously lifted the wheel. By Friday it was again lying on its temporary supports running the full length of the site, and its cables had been disconnected. Within a week it had been broken down into its container-size travelling components and sent on its way. With the final preparations completed, the swinging of the rim to vertical could now proceed. Achieved using the four permanent stay cables that run from the rear end of the spindle down to the fixing brackets at the bottom of the A-frame legs, it was a delicate operation, not least because the threaded connections at the end of the stays had to be carefully slotted into their respective brackets while the cables were under tension. Starting at 9 o'clock on the Saturday morning, the entire operation was to take more than 18 hours, the rim finally reaching the vertical position in the very early hours of Sunday morning.

By the beginning of November, British Airways London Eye was firmly established as a major new landmark on the city's skyline, as this panoramic photograph *(main spread)* taken from the roof of Millbank Tower shows. The London Eye is the capital's fourth highest structure – after the old NatWest and Post Office Towers and Canary Wharf – and can be seen from all over London, with clear vantage points as far distant as Hampstead Heath in the north and Richmond Park to the south-west. With the wheel held firmly in the vertical position, work had commenced immediately on commissioning the main drive systems *(left)*. All went smoothly and the rim made its first tentative revolutions during the last days of October, clearing the way for the arrival of the first capsules on site on Saturday 6 November.

From the outset, David Marks and Julia Barfield were determined that the capsules should combine unprecedented stability with the best possible views. Instead of the top-hung passenger cabins of traditional observation wheels, therefore, with their tendency to sway and support frames that obscure the view, they proposed that their capsules should be supported in two rings held well away from the rim, promoting uninterrupted views in every direction, and be fitted with an active drive mechanism that would keep the floor level at all times. It was a bold challenge that was to require the talents and expertise of a wide number of people. Freelance designer Nic Bailey joined the team in November 1996 and was instrumental in refining the initial concept, but the major breakthrough came with the appointment of the Poma Group of France, specialists in the design, fabrication and installation of all kinds of cable-car systems. Acting initially as design consultants only, the company was formally commissioned in September 1998.

The Capsules

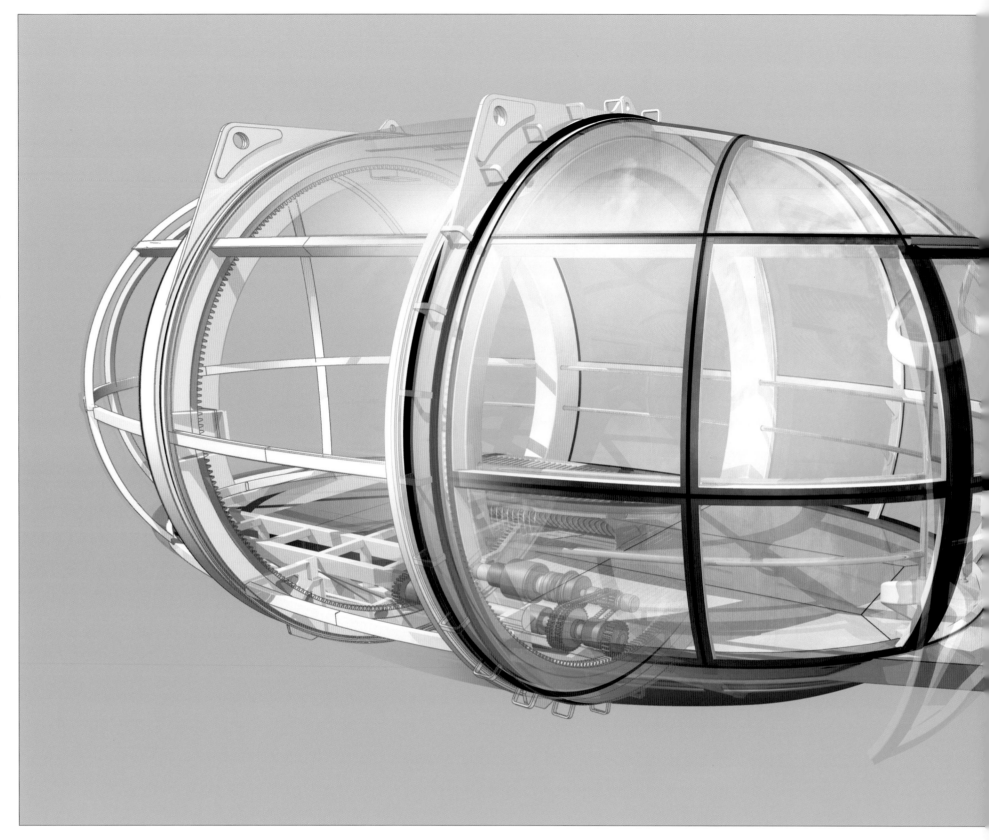

"I have known David and Julia for years, but we hadn't really spoken about the wheel until we met up by chance at a design show in London in October 1996. We talked about the work I was doing, the yachts and so on, and the 'Airstream' caravan I had recently redesigned. Anyway, something must have clicked, because a month or so later they asked me to help out with the design of the capsules. The general principles of the design – the ring supports, the American football shape, even the overall dimensions – had been fixed during the competition stage. My role, therefore, concentrated more on how the capsules might be built – identifying the most appropriate materials, exploring how the overall proportions might be adjusted to simplify fabrication, that sort of thing – but all the while respecting David and Julia's initial vision and trying to maintain the purity of the overall shape. Arup's had done some work on the mechanical systems and sketched out a basic structure but, even so, many of the practical issues were still open to discussion. We quickly reached the stage where it was sensible to seek further input from industry, from those companies that might actually manufacture the capsules, and we approached a number of companies we thought might be able to help – from boat builders and car makers to the manufacturers of railway carriages and monorail systems."

Nic Bailey
Design Consultant

The early design studies for the capsules — illustrated in freehand sketches *(below left)* — show a basic concept that has remained virtually unchanged. Even the overall dimensions are the same, although at this stage it was assumed that all passengers would be seated. Though Arup's made a series of practical studies, it was not until planning was approv-ed in April 1996 that the design of the capsules accelerated significantly, most notably with the addition to the team, in November of that year, of freelance designer Nic Bailey. With the skilled use of computer graphics — including the drawings of the final design shown here *(far left and below)* — Nic was able to refine the basic concept to the point where the Poma Group of France, working through its subsidiary Sigma SA, was able to produce an accurate mock-up *(bottom left)* in late 1997. Following the Group's formal appointment in September 1998, Sigma decided — in association with Semer SA, a sister company in the Poma Group with extensive experience in the design and production of complex electronic operating systems for all types of cable car systems — that the best way forward was to build a full-scale test-rig, in which some of the ideas outlined in the mock-up could be thoroughly assessed.

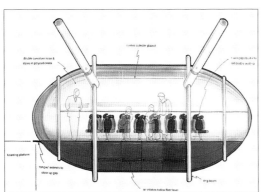

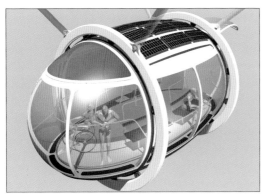

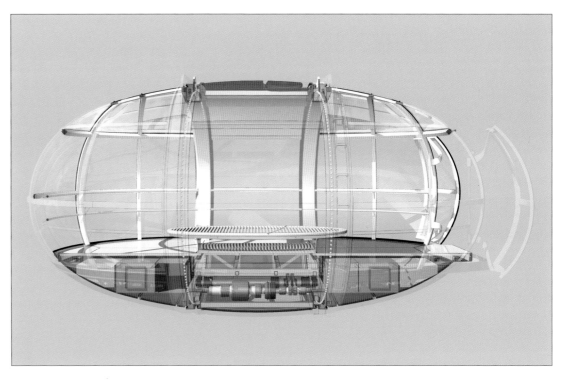

"I have been working with the Poma Group on an ad hoc basis for the past 17 years, ever since they installed one of their cable cars at my own theme park here in Derby-shire. I had seen the reports about the London Eye project in the trade press and realised that Poma's experience in the design and fabrication of both cable-car and monorail systems might be of relevance, and I contacted David and Julia to offer our services. It turned out they had already spent some months looking for companies they could work with to develop their concept. I organised a trip down to the south of France for David and Nic Bailey, where they met Jean-Paul Cathiard, the President of the Poma Group, and saw the factories there and examples of the company's work. They must have been impressed because, soon after, we were commissioned to make the full size mock-up."

Andrew Pugh
Managing Director, The Heights of Abraham

The design of the test-rig was a major undertaking in its own right. Sigma had to find a way of turning the capsule's external fixing points around the full diameter of the capsule, in such a way that simulated exactly the movement of a capsule around the finished wheel. After careful consideration, this was achieved by fixing the capsule to the inside of two large braced rings, supported on rollers. These external rings could then be rotated at the appropriate speed by an independent electric motor, while the drive mechanism in the capsule kept the capsule itself level. In its final form, the drive mechanism — which together with all its control and monitoring devices is now known as the stability system — has to respond to two basic conditions: it must synchronise the speed of each capsule's rotation to match exactly the turning of the rim, gauging its position on the wheel and holding the floor perfectly level, while at the same time responding to the changing loads imposed on the floor due to the movement of the passengers. In Semer's solution, the turning speed of the drive mechanism is constantly adjusted by a series of finely calibrated inclinometers, which are able to restrict the movement of the floor to less than 0.1 of a degree.

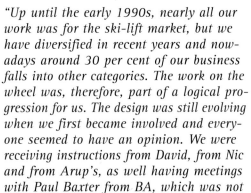

"Up until the early 1990s, nearly all our work was for the ski-lift market, but we have diversified in recent years and nowadays around 30 per cent of our business falls into other categories. The work on the wheel was, therefore, part of a logical progression for us. The design was still evolving when we first became involved and everyone seemed to have an opinion. We were receiving instructions from David, from Nic and from Arup's, as well having meetings with Paul Baxter from BA, which was not always easy. The real breakthrough came with the making of the mock-up — suddenly, everyone could grasp exactly what the capsule would look like; what it would feel like inside. It also helped resolve a number of key issues. We studied loading procedures, for example, by pulling the mock-up past a platform at various speeds to see how many people could get on or off in set distances."

Jean-Francois Savard
Managing Director, Sigma SA

"In essence, when we were approached again in September 1998, we had to start from scratch. Just how the mounting rings and big bearings might work had been discussed, of course, during the making of the mock-up, and we had some ideas for the main structure, but no calculations had ever been made. We had even had some trial pieces of glass made by a Swiss company, to see what we might learn about double-curved glass, but these only demonstrated how difficult it would be. Even so, we were fairly confident we could make our basic concept work. Our main concern was the stability system. Keeping the floor level was not a problem if the loads were evenly balanced, but we had no idea how the system would react as people moved around. Would you feel any movement? If so, would it be acceptable, or was it better to remove all sensation of movement entirely? We couldn't be sure and it soon became clear that most of these issues could only be resolved using a full-scale test-rig. Unfortunately, there was very little time. Many of the components – the main ring bearings, for example – had to be ordered well in advance, so we ended up making the test-rig with the real pieces as they became available, which was not ideal. Subsequent changes might be possible to some elements, but not to others. It was a huge relief, therefore, when the stability system worked perfectly from its first run."

Philippe Desflammes
Project Manager, Sigma SA

155

"Curiously, the overall dimensions of the capsules have not changed that much. We did look at what savings might be made if the capsules were smaller, but they turned out to be quite minor. You saved on materials, but the number of joints remained the same and, if anything, the overall complexity only increased, as it became ever more difficult to fit all the equipment in."

Nic Bailey
Design Consultant

The expense of complex tooling could not be justified for 32 capsules alone and, during the early stages, much of the basic structural frame was made by hand, the work being undertaken by skilled welders at the Sacmi factory, a sister company of Sigma in the Poma Group. As with the rim, though on a far smaller scale, the separate elements — including the floor *(below)* and the curved glazing frames that form each end of the capsule *(left)* — were formed in pre-cut and, in the case of the latter, pre-curved steel sections assembled on master jigs. With the rapid heating and cooling caused by the welding process distorting the metal slightly on each occasion, a strict sequence had to be maintained to ensure subsequent welds 'balanced' each other out.

"From the outset, the real problem was always going to be meeting the delivery dates. There was very little time to spare, so it was important everything ran smoothly. There was a great deal of pressure to complete the design and confirm the calculations as soon as possible, so that orders could be placed in good time. Due to the nature of our work, we rely on subcontractors a great deal and we had to work to their schedules. Fortunately, we developed a very good relationship with Allott & Lomax, which allowed decisions to be agreed quickly. We could keep everything moving forward, but this only emphasised that there was no time to change your mind, let alone start again. There was always the slight doubt that maybe there was a better solution. Fortunately, we knew the mechanical systems we had worked very well and there was no need to change them. This was not always the case, however, with those issues affecting the 'look' of the capsule. David and Nic had a specific vision, a particular feeling for what the capsules should be like. It was our responsibility to turn that dream into reality — into actual shop drawings, calculations and materials — but finding a solution that everyone was happy with could be very difficult, and on many occasions the schedule was stretched to the absolute limit."

Jean-Francois Savard
Managing Director, Sigma SA

With insufficient space within its own factory to build 32 capsules, Sigma decided to rent and recommission a vacant factory in Grenoble which was almost purpose-made for the job. Full-height access at each end of the long space, which came complete with travelling cranes in the roof, allowed a clear production line to be established, with individual components arriving at one end and finished capsules leaving at the other. Taken over in late April 1999, the assembly of the first 12 capsules was well under way by the time these photographs were taken in mid-June. The basic structural components could hardly have been simpler. First came the fully braced floor frame into which was installed, at this stage, only the drive mechanism for the stability system *(below)*. Once this had been completed, the floor was then inserted into what is effectively the capsule's chassis: a central structural 'drum' that supports, at each end, the four-metre diameter ring bearings *(right)*. The floor frame, the central drum and the end glazing frames are all supported on the 'internal' side of each bearing, while the 'external' side incorporates the pin-joint connections that will attach the capsule to the rim. A toothed rail fixed to the 'external' side of the bearing provides the purchase against which the cogs of the capsule's drive mechanism can turn, and thus maintain the floor at horizontal.

All of the capsules' operating systems — including the lighting air conditioning, public address, radio links and emergency sensors, as well as all the controls and monitors for the stability system — were developed by Semer at their factory near Chamonix, at the foot of Mont Blanc, and installed by them during the capsules' assembly in Grenoble. To ensure the systems run smoothly and without interruption at all times, each capsule is fitted with duplicate control boxes, located at each end of the floor space, one acting as a permanent back-up to the other. Both boxes can draw power from one of two separate power supplies, while back-up batteries hold sufficient power to run essential services for up to three hours in the event of an unexpected power cut.

"There is an awful lot of mechanical equipment in the floor and we spent a great deal of time trying to find the best way to fit it all in. We didn't want bulges appearing in the external skin, but at the same time we didn't want the floor rising too high into the capsule and reducing the sense of spaciousness that both David and I always saw as an essential part of the ride's overall quality."

Nic Bailey
Design Consultant

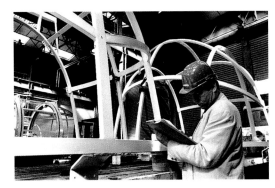

"Semer have a long experience of designing automatic control systems for all sorts of industrial and public transport systems, from quarries and steelworks to automatic monorails systems and every type of cable-car operation. All of these applications are safety critical in one way or another, so we are used to designing our systems to very strict specifications and with high degrees of redundancy. We have to consider the hardware, of course, and make sure it will operate reliably under the expected conditions, but more importantly we must develop good software – compatible with the hardware and easy to use, both in operating and monitoring mode. The biggest danger in any complex system is the operator misunderstanding the information. To ensure the best possible standards, we have developed our own quality plan, running two software design teams against each other, one to design the system and the other to test it, to expose its flaws. Improvements are made, then tested again until we can be sure any bugs have been removed. Much of this comes down to instilling a culture of safety in the workforce. In fact, a high proportion of our people come from the aeronautic or transport industries, where safety is already a priority. After that it is a matter of checking and testing at every stage: in the laboratory, in the factory and finally, of course, on site."

Marc Bottelier
Project Manager, Semer SA

"On a project of this scale and complexity you are always looking for a safe pair of hands. This was especially the case with the glass – I must have spent around three months chasing around Europe trying to identify the right manufacturer. Finding the quality we required at a reasonable price proved to be a major challenge. In the end, we invited three companies to prepare samples for us – one from Britain, one from Spain and, of course, Sunglass from Italy.

Quite frankly, there was no competition and Sunglass got the job. I undertook a number of these company reviews for all sorts of components, and it was always an eye-opening experience. With such a high profile it was hardly surprising that the project was so attractive to many companies. What was odd was that they all seemed to have an opinion – even when, as was often the case, it soon became clear they had no idea what the job actually entailed. There were occasions when you could only assume

somebody must be having you on. It was such a relief to come across companies that had an immediate grasp of the technical and programming difficulties, and could offer sensible solutions."

Neil Thompson
Design Manager, Mace Limited

The capsule glazing had to be manufactured to very exacting standards to maintain both the perfectly smooth shape the designers had in mind and the highest optical quality — not always easy when every panel is curved in both directions, in the case of the main end panels to a degree that had never previously been attempted. The Italian manufacturer Sunglass s.r.l. was identified as one of the few companies capable of completing the job to the quality required, and the work was carried out at its factory near Padua. For the simpler glass panels, two layers of 6mm-thick glass — precut to the correct size — were placed in specially produced moulds into which the glass 'fell' as it was heated. The two layers were moulded at the same time to ensure a perfect 'fit' during the lamination process. For the more demanding end panels, with their very deep curvatures, Sunglass developed and refined their own techniques to ensure every panel was produced to the requisite optical quality.

As each glass panel was completed, it was checked on a specially fabricated jig (left) to ensure it matched the required shape exactly.

"Glass technology has improved dramatically in recent years. There have been many developments in the car industry, for example, but the shapes still tend to be quite simple. Curves are either relatively shallow or are more acute in one direction only. It soon became clear that we needed to find a company willing to work well beyond the usual standards."

Nic Bailey
Design Consultant

The lamination process began with the thorough cleaning of the two glass sheets that make up an individual panel *(right)*, after which 1.5mm of lamination film, in one or two layers depending on requirements, was stretched between them and taped into place *(far right)*. The completed 'sandwich' was then sealed in a vacuum bag *(opposite)* and carefully stacked on a purpose-designed trolley, ready for transferring to an autoclave. Here, the panels were reheated under vacuum, following a very precise heating and cooling programme, causing the lamination film — opaque in its initial form — to clear and bond completely with the adjacent glass surfaces. Absolute precision was essential as the slightest flaw can cause the panel to crack.

"Sigma first contacted us in mid-September 1998, though at that time this was only to see if we were interested in the project and to check if we thought it was actually possible. There were many technical complications, but we liked the challenge and that led to us being asked — in competition with two other companies — to make up some test panels. The tolerances were very tight indeed. Because of the way the glass is fitted, without external frames, the individual panels had to be cut to size with great accuracy and shaped precisely. With 36 panels of glass making up just one capsule, this was already a significant challenge. And because of the capsule's elliptical floor-plan, the curves are not always constant. It turned out that there were seven distinct shapes for bending, six of which worked in symmetry — requiring 13 moulds in all. The real challenge however, was maintaining a high optical quality in the two main end panels. As a rule of thumb, we normally

Sunglass provides a complete service, including its own lamination process which is carried out in a dedicated 'clean' room, fitted with its own climate control to ensure optimum, dust-free working conditions at all times.

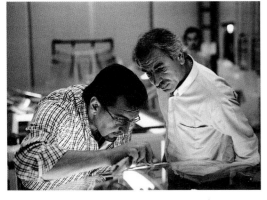

work to a maximum camber of around one in eight — that is, the height of the curve above the plane formed by the panel's sides is never more than one-eighth the length of the shortest side. Here, we were being asked to produce panels of glass with cambers of up to one in four — and not just single sheets, but always two sheets perfectly matched for lamination, which requires that the touching surfaces are exactly parallel. There was a lot of research — different types of glass, various temperature gradients and so on — and we made at least a dozen prototypes, recording and analysing the results at every stage, before we achieved the required quality. The final challenge, after that, was to achieve the same results in all 1152 panels."

Giuseppe Bergamin
Managing Director, Sunglass s.r.l.

Weighing anything up to 80 kilos each, the 36 glass panels that make up the glazing for one capsule had to be fitted with extreme precision to ensure the seamlessly smooth form specified by the design. The upper panels were positioned first with the help of a hydraulic lifting bracket *(below)*, being carefully aligned on the white-painted steel frame with the use of shims and spacers in the form of continuous strips and blocks. Once their final position had been approved, the panels were then fixed in place with black silicon sealant, applied in two phases: the main fixing layer first, injected internally between the glass surface and the steel frame *(below left)*, followed by the weather seal, injected externally between the panels. A back-up mechanical fixing is also provided, in the form of a stainless steel disc, fixed back to the steel frame at the point where the corners of four panels come together. Unsurprisingly, it did not take long for someone to notice that the complex ellipsoidal shape of the end glazing frames bore a remarkable similarity to old-fashioned 'pen-tops'. The name caught on and was used thereafter, in France as well as the UK.

"We did look at the possibility of using tinted or mirror-coated glass for the direct vision panels – the roof glazing is coated slightly – but I always felt that clear glass was the right option in thermal engineering terms. This is because the application of any form of treatment causes the glass to heat up when exposed to the sun and, in my experience, this heat is radiated into the interior in a way that is more uncomfortable than the heat of sunlight felt through clear glass. Furthermore, clear glass allows much of the incident solar energy to pass right through the capsule. This goes rather against conventional wisdom, but I firmly believe that, in this particular application, it is appropriate. Provided the capsule air temperature is kept at the right level, passengers can then move around freely – in and out of the sun – until they find the balance that suits them best."

Loren Butt
Director, Loren Butt Consultancies

"One of the great benefits of having all 32 capsules under the same roof, and at different stages of construction, was that we could adapt very quickly to any necessary changes in the programme – when particular components didn't arrive on time, for example – allowing us to use the time available to the best possible advantage."

Andrew Pugh
Managing Director, The Heights of Abraham

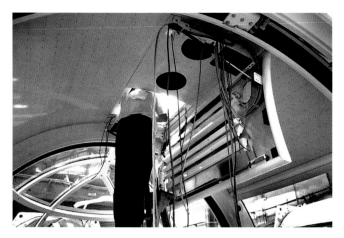

The fit-out of all 32 capsules was a long and complex procedure, completed in just over five months between late April and early October 1999. For the most part, they were prepared in sets of four or six, in recognition of the weekly transport schedules that came into effect when the first capsules — complete in every detail except for the central wooden bench — left the Grenoble factory at the beginning of September

(above). Working upwards, the fit-out of the systems within the floor frame was completed first, followed by the fitting of the floor panels. With the floor in place, the fixing of the final wiring around the central drum and behind the ceiling panel could be completed with relative ease. With the glazing in place, all that remained was the fixing, both internally and externally, of the fibre-glass cover panels and, of course, the doors.

"One of the hardest things for us to take on board was the realisation that there was no comparison between the capsules and a standard gondola. We had to eliminate a whole way of thinking. The primary purpose of the typical gondola is to move the maximum number of people in the shortest possible time. Skiers do not normally worry about the view and they do not expect high standards of comfort; no heating is required and lighting is normally minimal. In normal operations, a ski-lift is little more than a Metro system in the air. With the capsules, on the other hand, the quality of the ride, the comfort of the passengers, is everything. Sudden or unexpected movements must be eliminated; there has to be heating and cooling available at all times, and the best possible finishes are expected. A gondola of this size would probably carry more than 100 people, but would weigh no more than 1.5 tonnes when empty. The glass used in a single capsule weighs that amount on its own, while the stability system adds a further tonne if the bearings are included — and all this, of course, requires a heavier and stronger structure to hold it in place. It is hardly surprising, therefore, that a capsule designed to carry 25 adults has ended up weighing around 10 tonnes."

Philippe Desflammes
Project Manager, Sigma SA

It had been understood since the competition stage that the capsules, wherever they were made, would have to make some of their journey to site by road and this, in part, determined that their overall diameter was fixed at around 4 metres — well within the maximum width allowed to travel on public roads without a full police escort. What this calculation failed to take into account, however, was that at this diameter, even when loaded on the lowest flatbed trailer, the finished capsules would stand some 4.8 metres high. This is fine in the UK, where most motorway bridges have a clear headroom of 5 metres or more, but was not a great help in France where some bridges across the autoroutes have been designed with allowable headrooms as low as 4.3 metres. There was no alternative but to find a route from Grenoble to Zeebrugge that avoided all low bridges. During the seven-week transport period that followed, villages the length of France became used to the sight of yet another *'convoi exceptionnel'* passing through their midsts *(below and far right)*, as the capsules made their long journey to the English Channel — and from there to a secure holding area near Dartford Bridge *(below right and bottom)*.

"We were first approached by Sigma some time in March 1999. We had not worked with them before, but as one of Europe's leading specialists in the transport of oversized or delicate loads, it was not that unusual to be approached by a French company. I must say they had an interesting problem for us. At around 10 tonnes, the capsules weigh considerably less than a standard container, so the weight was not an issue. There is a lot of glass, of course, which is always a concern, but the real

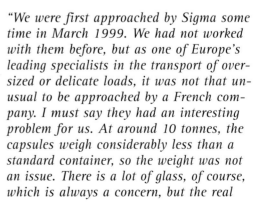

problem here was always going to be the overall dimensions. At four metres wide, a full police escort would not be required — although they did prove necessary at certain localised points — but there were all sorts of other arrangements to be made. Permits had to be approved; English and French pilot cars and their crews had to be hired; appropriate ferries had to be booked; and, most importantly of all, a route had to be found that avoided all obstacles. And the key obstacles in France, we knew from experience, would be the fixed road bridges which are generally very low. We knew Zeebrugge in Belgium would be the destination, as that is one of only two ports operating ferries large enough. The trick was sorting out what happened in the 500 miles in between. A certain amount could be achieved with maps, but at a certain point there was no alternative but to travel the route by car. Every thoroughfare and junction had to be considered, starting with the decision to turn left or right out of the factory. With* *that out of the way, it was a question of logistics, with every town and locality having to be approached to see what permissions might be required, and not just once but for 16 separate convoys, each of two capsules, travelling at three or four day intervals over a seven-week period. After all that, the transfer in England to the holding point at Dartford was relatively straightforward."*

Peter Dafters
Managing Director, Peter Dafters (STG) Ltd

Under pressure to make up the time lost during the delayed up-lift, the fitting of the capsules proceeded very quickly indeed. By working late into the night, all 32 capsules were fitted in just eight days.

To facilitate the lifting of the capsules, a purpose-made platform and mobile lifting rig was positioned directly under the wheel at the same level as the boarding platform. This projected out into the river sufficiently to allow one capsule at a time to be lifted directly from one of the two barges (right), specially converted to transport the capsules upriver from Dartford, into the mobile lifting rig (above). The rig was then pulled back into position below the rim and aligned with the support frame above. Four hydraulic jacks — one at each corner — were then able to raise the capsule into place, the jacks operating independently to allow the pin-joints to be aligned precisely.

It had been realised during the earliest stages of the structural design that the fitting of the capsules to the wheel would be a critical operation. Projected away from the rim, the connection between the capsule and the support frame acts as a cantilever in which the direction of the forces is constantly changing as the wheel rotates. The most efficient connection in these circumstances is a pin-joint, but these must be fitted within very tight tolerances if they are to work properly. With the capsules made in France and the rim in Holland, careful co-ordination had been essential throughout the fabrication process to ensure the two would come together exactly in London. It is a credit to the teamwork that evolved on the project that all went according to plan and the first capsule slotted into place at the first attempt.

"We had carried out extensive tests of the capsule systems all the way through the fabrication phase, of course, in the factory and on the test-rig. The test-rig had been especially useful, letting us run the stability system for extended periods — in fact we were still trying things out there when the real commissioning started on site. But nearly all of those tests had involved only the capsules' internal systems, tested one at a time. Everything changed when we got to site. For the first time we could control and monitor every system for all 32 capsules simultaneously. It was a painstaking operation. Every system had to be tested again individually, partly to ensure they were all fully operational, but also — and more importantly — to ensure they were all feeding back accurate information to the control room on the boarding platform. Everything had to be checked and rechecked."

Steve Saint-Soupane
Installation Manger, Semer SA

The last capsule was fitted to the wheel on Sunday 14 November and because connection to the rim's main power supply had been made as each capsule was lifted into place, it only remained for a few safety checks to be made and the power switched on for the final commissioning of the capsules to begin. Taking into account the number of sensors in each capsule – for temperature, lighting, electrical supply, battery back-up, clutch, drive motor, the doors, whether they are open or closed, locked or unlocked, for the escape hatch, for the inclinometers, smoke detectors and so on, many of which are duplicated for safety reasons – as well as the interaction required with other contractors, this was an arduous and time-consuming activity that would take the best part of three months to complete.

"Communication between the control room and capsules is carried out by radio. We had considered running a cable link to each capsule, using a bus-bar system similar to that used to provide power to the rim, but we were concerned that possible breaks in the connection might corrupt the flow of information. That could happen with radio links too, of course, but we have guarded against that by using a spread-spectrum system where several wavebands are monitored simultaneously, allowing the information to be transferred automatically via the strongest signal. These operate at frequencies significantly different from those used by other communication systems, and the transmitters are highly directional, so there is virtually no chance of interference. With so many sensors on each capsule, we are talking about an enormous amount of information. Fortunately, we have installed a number of large networks, so although much of this work was very time-consuming, it was relatively straightforward. We were actually far more interested in the close liaison we needed to establish with other companies. We usually work on our own projects, but here it was necessary to build good working relationships and to find technical solutions that were not only good for ourselves but also linked well with others. In fact, the level of co-operation on site was very good, and the work went very well."

Marc Bottelier
Project Manager, Semer SA

In the run up to Christmas and, indeed, well into January – in fact, until all the required safety checks had been approved and signed off – the wheel was meant to turn, other than for key personnel, with the capsules empty. In reality, the temptation proved just too great and key members of the design and construction team enjoyed a few 'trial runs'.

"You couldn't help but get caught up in it. I remember I was flying back from Holland on the day the first capsule was lifted into place. The flight landed late in the evening but there was no thought of going home – I just had to come down to site and see how it had gone. I knew I wouldn't have been able to sleep until I'd seen the pins in place, seen what the tolerances were."

Neil Thompson
Design Manager, Mace Limited

175

By late November, the wheel may have been up and even running, but there was still a considerable amount of work to do. First and foremost, the last of the temporary river platforms, including the one installed to facilitate the lifting of the capsules, had to be dismantled so that the large construction barges could be withdrawn. Only then could work begin on the installation of the ferry pier and collision protection system. One of the contractors under greatest pressure during this period was T. Clarke plc, the company with responsibility for the installation of the project's electrical supply network. It was essential that power be provided to the wheel as quickly as possible if commissioning of the capsules, the drive mechanism and other operating systems was to progress smoothly. But all the following fit-out and finishing trades – including the landscape contractors – were under the same kind of pressure, all playing their part to ensure British Airways London Eye was ready to open to the public in February 2000 as scheduled.

Planning guidelines have limited the number of permanent on-site structures to two timber-faced kiosks, which were delivered to site in prefabricated sections *(above)*: one to operate as a snack-bar and coffee outlet, the other as a shop.

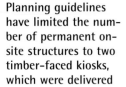

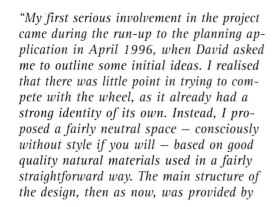

"My first serious involvement in the project came during the run-up to the planning application in April 1996, when David asked me to outline some initial ideas. I realised that there was little point in trying to compete with the wheel, as it already had a strong identity of its own. Instead, I proposed a fairly neutral space — consciously without style if you will — based on good quality natural materials used in a fairly straightforward way. The main structure of the design, then as now, was provided by an avenue of trees running the length of the site on axis with the wheel. This creates a formal approach for those arriving from that direction, while at the same time defining more relaxed areas on either side. My main concern, at all times, was that as much of the site as possible should be seen as a new public urban space, capable of embracing a range of activities."

Edward Hutchison
Landscape Architect

Aware that work would have to continue on site well into the New Year, it had been decided that priority for the landscaping would be given only to the pedestrian bridge linking the site to a new side entrance into County Hall *(below)* and to those areas around the boarding platform and kiosks. The rest of the site would have to be completed at a later date. Following the installation of the underground services — not least the main power cables connecting the project's own 11,000-volt underground substation near Belvedere Road to the main distribution switchrooms on the boarding platform — work began immediately with the laying of the granite sets and the fabrication of the reinforced concrete bench that forms the boundary between the site and Jubilee Gardens.

"I think it is fair to say that the amount of work we undertook between September '98 and December '99 was far greater than any of us had ever envisaged. I was in the fortunate – or maybe unfortunate – position of being able to assess nearly every aspect of the project and, for whatever reason, there is no disguising the fact that it was never really as advanced as we thought it was. All the permissions we needed before work could commence on site, the various issues with the local authorities, the engineering challenges, the overall complexity of getting all the different pieces to site at the right time – nearly every aspect when looked at in detail – was far harder to resolve and far more time-consuming than we had ever thought. Quite frankly, I was amazed, come December '99, that we were still there or thereabouts with the programme. I believe a good many of us will look back on this project as one of the highlights of our professional lives, in part because of the unique challenge it provided, but also because of the intensity of the whole experience, the intensity of the relationships we established and the intensity of the camaraderie we all enjoyed. Despite the odd glitch, the odd embarrassment – of which there were always going to be a few – the construction has been a resounding success. And that, in itself, is a testament to the unreserved dedication and perseverance of the whole team."

Tim Renwick
Project Director, Mace Limited

Throughout November and December, while the landscaping work continued at ground level, a considerable amount of final commissioning and testing was being carried out on the wheel itself, both within the hub and spindle assembly – approached via access ladders inside the A-frame legs – and out on the rim. Specially trained in all forms of climbing, abseiling and working from ropes, much of this work was undertaken by operatives from CAN Limited. All went well and by the week before Christmas, preparations for the official switching on of the wheel on New Year's Eve were well advanced. The ferry pier and its access bridges were now in place and the permanent lighting for the whole site was up and running. With a stately rim speed of just 0.26 metres per second – or a little over half a mile per hour – when the wheel is turning at its normal operating speed, movement of the capsules is almost indiscernible during the day, and only shows up on this night view from the Royal Horseguards Hotel (below), on the far side of the river, because of the photograph's long time exposure. The sunset view (left) from the nearby Shell Building, on the other hand, was taken when the wheel was stationary, allowing the rim to stand out in sharp silhouette against the western sky.

Final fit-out of the ferry pier pontoon was carried out at Tilbury Docks *(far left and below)*, while the floating booms *(left)* were fabricated at Tilbury Douglas's own works at Belvedere.

"Considering all the difficulties of the site – the limited access, working over water, too many people in too small a space – it is all the more impressive that we avoided any serious injuries. And as far as I know, nobody fell in the river. Actually, that's not quite true. I did see one person slip and half fall in, but he was pulled out quickly so I'm not sure that really counts."

Jan Stam
Site Operations Manager, Hollandia BV

Conceived by the Beckett Rankine Partnership and developed through to detail design by Tony Gee & Partners, with specialist support from Houlder Offshore Engineering, the independent ferry pier is actually part of the wheel's collision protection system. The pier itself — actually a floating pontoon — its access bridges and the floating booms, that stretch at an angle from both ends of the pier to the river wall, act as a unified system, capable of halting a 2200-tonne vessel travelling at 8 knots — this being the largest vessel that uses the Thames on a regular basis. With time at a premium, Tilbury Douglas subcontracted the fabrication of the pontoon to Wear Engineering, while the access bridges were built by their sister company Nusteel and fitted out by Littlehampton Engineering.

"Our first involvement with the project came about when a couple of boat operators we regularly work with mentioned they thought there should be a ferry pier associated with it. From the drawings we had seen – this was around October '97 – we knew there was a fairly straightforward boom arrangement that acted as a collision protection system, but there was no provision for landing boats. Working speculatively, we came up with a scheme which replaced that design with a floating pontoon, held out in the river by a pair of access bridges (or brows) that act as radial arms, with the collision protection booms now held at each end. By connecting the booms to a series of energy-absorbing units manufactured by Jarret of France – similar to hydraulic pistons but filled with silicon putty instead of oil – we calculated there should be enough elastic deformation in the system to stop even the largest boats. Of course, this would be more expensive, but it should have been possible to recoup the costs from the boat operators. As it turned out, when Marks Barfield Architects asked us to develop the concept, we found out that the Millennium Commission was willing to find half the cost themselves, as part of their programme to promote transport on the river. In the end, of course, we became involved in the project in another capacity and the scheme was worked up by others, but it is still nice to know we played our part."

Tim Beckett
Partner, The Beckett Rankine Partnership

Midnight on New Year's Eve, and the 'River of Light' explodes over British Airways London Eye to welcome in the new millennium. It had been accepted some months before that any chance of opening to the public on 1 January 2000 was just not on – it had always been asking a great deal to expect that what had initially been conceived as a two year construction programme could be completed in just 15 months. Even so, it had seemed that sufficient capsules would be ready to allow a limited number of private guests to enjoy a ride on this special night. And all seemed to be going well until, quite unexpectedly on the day before New Year's Eve, a minor fault was detected in the clutch on one of the capsule stability systems. An identical clutch had been running perfectly on Sigma's test-rig in France for more than eight months, but with no time to strip the clutch down and check what the problem might be, there was no alternative but to run the wheel without passengers. It was a bitter disappointment, but everyone agreed that safety was paramount. Subsequent analysis found it to be a rogue manufacturing fault that affected only a handful of clutches, but as an extra precaution all were replaced before the wheel opened to the public.

From the beginning of January, while the final fit-out of the boarding platform continued *(right and below left)*, key personnel from British Airways London Eye's own technical and operating staff — who had already undertaken extensive training off-site — started the lengthy procedure of acclimatising themselves to the working conditions on the wheel itself, under the watchful eye of Operations Manager Paul Kelly and Technical Operations Manager Kevin Dyer *(below right, top and centre)*.

"I have been involved with the project since July '99, liaising with the manufacturers to gain a complete understanding of all the systems, then deciding what size team was required and arranging their training. We have ended up with 17 engineers in all — all with backgrounds in fields where public safety is a priority — who work round the clock in shifts, with three on duty during the day and five on call for overnight maintenance and testing. This is a lot heavier than usual, mainly because the critical parts tend to be those that carry passengers. On a normal ride that usually means no more than one or two trains, but here, of course, there are 32 capsules, each of which has to be checked individually. Every morning, before the wheel opens to the public, each aspect of the wheel that is safety critical or will affect passenger comfort is checked, either visually or more often via the computer, while some — the door locks, for example — are tested manually. It is a time-consuming operation that can take the night crew anything up to four hours to complete, over and above any other work that might have been necessary. And, of course, many of the systems are unique, so we couldn't be entirely sure how they would perform in practice. More than a few have required constant monitoring and only now, after five months of running, are we settling into a steady routine."

Kevin Dyer
Technical Operations Manager
British Airways London Eye

Extensive checks of all safety-critical systems — including the main drive mechanism (below) — are carried out every morning before the wheel is opened to the public

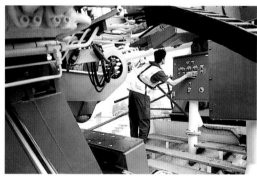

"There is so much information coming back from each capsule that we decided early on that all of the systems would have to be self-monitoring to one degree or another. In normal operations, there are two displays available to the operator: one showing the whole wheel and indicating the position and operational status of each capsule; and a second that focuses on each capsule in turn, showing both the key information — temperature, attitude and so on — and which systems are running. If a fault is located, the appropriate display shows up immediately on the monitor, the part in question flagged by colour. If the fault is minor, the computer will offer the operator a choice of options; if serious it will initiate the appropriate action automatically, stopping the wheel completely if necessary. Further layers of information can then be accessed by the technical staff, who will decide the next course of action."

Marc Bottollier
Project Manager, Semer SA

In its final form, the site landscaping can be identified as two separate zones: with hard-wearing granite sets at the top of the site, laid out to a strict grid around the two kiosks and towards the boarding platform, where visitors will naturally congregate; and a more relaxed area on the approach from Belvedere Road, dominated by the avenue of semi-mature double-white flowering cherry trees *(opposite)* and open squares of Cotswold gravel separated by granite steps *(below)*. A bed of ground-cover bamboo and evergreen grasses protects the North Retaining Wall — a listed structure next to County Hall — and complements the 10-metre high trees. Some 15 years old, the trees could well grow to a height of 60 metres over the next 50 years. Creating a *cordon sanitaire* around the wheel, the security features have been cleverly disguised as public seating: the wall separating the site from Jubilee Gardens as a granite-faced bench topped by a hedge, and the individual traffic bollards as solid granite seats, 400mm square and 600mm high.

"The curious thing is that the extreme pressure of building such a high profile project helped make it a success. Because everyone knew this was a special opportunity, like building a new Eiffel Tower, they were willing to find the extra time and make sure they gave the best advice. The best people wanted to be involved and they too were willing to give that extra 10 per cent. This is reflected in the accuracy of the rim. Privately, we had assumed that the best construction tolerances we could hope for, in terms of the rim being out of plane, would probably be around ±100mm — although we had actually allowed for more than twice that amount. In fact, the final survey has confirmed that no point on the rim is out of alignment by more than 40mm. To put that in context, if the wheel was scaled down to the size of a bicycle wheel, the rim would be out of true by less than one-third of a millimetre."

Professor Jacques Berenbak
Design Engineer, Hollandia BV

To allow the operating staff at least a few quiet weeks to settle into their jobs and establish the routines necessary to keep the wheel running smoothly at all times, British Airways London Eye received its first paying customers without fanfare from the beginning of February 2000, with ticket sales limited to keep numbers to a manageable level. Having proved all was well, the official opening was then held on a rather grey Thursday morning on 9 March, in the presence of over 1000 assembled guests and team members, with the rest of the day given over to rides for the press and the media. The wheel had already proved its popularity during February, but the crowds that flocked to London's newest attraction over that first weekend after the opening *(below)* took everyone by surprise. Inevitably, its very success exacerbated the few teething troubles there were, but an active programme of improvements and additions over the following months meant everything was running smoothly by the summer months *(left)*.

If the fine-tuning of the capsules took a little longer than planned, the main drive and stability systems on the restraint towers *(opposite)* have run almost perfectly from day one. The effortless ability of the hydraulic drive to stop and start the wheel so smoothly has proved a particular revelation. During normal operations, the capsules can be brought to a complete stop in just two metres with no discernible movement of any kind. Indeed, the deceleration profile for the hydraulics has been refined so precisely that emergency stops within half a metre can be effected with absolute safety. As with nearly all the systems on the wheel, there are two entirely independent electrical supplies, both to the drive system and to the capsules on the rim *(below)*, controlled via the two main switchrooms on the boarding platform, each of which is capable of supplying all the main systems independently. A back-up generator, situated in the substation, cuts in automatically in the event of a power cut, with sufficient power to operate all essential safety services.

Both the main drive and lateral stability systems are mounted on heavy pin-jointed frames, with powerful rams that allow the different elements to be fully withdrawn for maintenance purposes.

"There was a lot of discussion in the early days about whether the capsules should be air-conditioned at all, and we did look at the options, but it quickly became clear that the capsules had to be enclosed — and that being the case, there was little alternative. With that decision made, what was imperative was to keep the energy demands as low as possible by specifying the most efficient systems wherever we could. With a power input of around 12 kilowatts each, one can't escape the fact that the capsules are inten-sively serviced, but this has to be balanced against the fact that this benefits up to 25 people. An office of similar floor area might require a third of the power input, but would benefit no more than two or three. The reality is that a small family will probably expend more energy driving the few miles to their local station than they will use on the wheel."

Loren Butt
Managing Director, Loren Butt Consultancies

With 32 capsules, each capable of carrying 25 passengers, the wheel has a maximum capacity of 1600 people per hour — though in practice ticket sales are normally limited to around 1400 to allow for the inevitable delays that occur in operations of this complexity. Open between 10am and 6pm during the winter months — from October to March — and from 9am to 10pm during the summer, target estimates had initially assumed peak capacities of around 14,000 people per day during the busiest periods, with markedly quieter days in between. In reality, the quieter days have been few and far between — even the rain, it seems, has done little to dampen enthusiasm — and over the first full year of operation British Airways London Eye averaged nearly 11,000 people per day. Indeed, on many occasions, daily throughput has reached well over 17,000 — exceeding the predicted 100 per cent occupancy rate by a considerable margin.

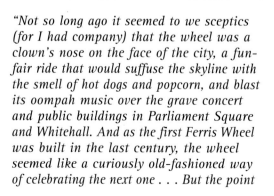

"Not so long ago it seemed to we sceptics (for I had company) that the wheel was a clown's nose on the face of the city, a fun-fair ride that would suffuse the skyline with the smell of hot dogs and popcorn, and blast its oompah music over the grave concert and public buildings in Parliament Square and Whitehall. And as the first Ferris Wheel was built in the last century, the wheel seemed like a curiously old-fashioned way of celebrating the next one . . . But the point the sceptics missed was that the sheer energy of the thing, and the simple joys it offers, outweigh the negatives. And we need not have worried so much about the wheel's threat to the dignity of Parliament, as the architecture of Barry and Pugin is not such a fragile flower as to be overwhelmed by a big circle of white-painted steel."

Rowan Moore
The Evening Standard, 18 October 1999

"Originally, we were planning for a permanent staff of around 75, with extra help brought in seasonally as required. Based on our experience with other attractions, you can normally predict when the busiest times will be — summer weekends, bank holidays, Easter and so on — and plan accordingly. But here, we have been running at capacity from the very beginning, with no 'down' time at all. I can't deny that this caused a whole raft of problems, and it made for a very chaotic first few months. Of course, it was better to be managing success rather than failure, but it was not ideal and I am only grateful that most of our visitors were so patient. Thankfully, the right procedures soon fell into place and we were able to take on the staff and management personnel we required — a full-time staff of 225, split into three shifts, that keeps things running smoothly, even on the busiest of days. It was a very steep learning curve but, quite frankly, I would not have missed it for the world. I am a lot older, I'm a lot greyer, but to be part of a project loved by all has been a fabulous experience. Everyone, it seems, wants to have a go. Age is not a barrier. Social class is not a barrier. It really is open to all. One only has to see the list of celebrities, politicians, ambassadors and even royalty who have visited over the past year to realise what an extraordinary impact the wheel has had."

Paul Kelly
Operations Manager, British Airways London Eye

By April 2000, just one month after the wheel's official opening and a mere six months since its uplift in the previous October, the gentle curve of British Airways London Eye arching across the skyline of the city had already become a familiar landmark, visible from all over central London — including St James Park *(left)* — and beyond. From closer vantage points, in particular from the Embankment on the far side of the river, the true scale and grandeur of the wheel is immediately apparent, as in this view *(main spread)* taken at dawn, with the top-most capsules and the towers of the Houses of Parliament glinting in the first of the day's sunshine.

As Henry Mayhew noted after an ascent in a hot-air balloon as long ago as 1882 *(see page 10)*, "there is an innate desire in all men to view the earth and its cities from exceeding high places", a fact confirmed by the immense popularity enjoyed by British Airways London Eye since it first opened. On fair days, and even foul, the world's tallest observation wheel offers unrivalled and ever-changing views. Indeed, one of its major successes has been the number of times visitors have returned to enjoy London's many different aspects. On a clear day the view can be breathtaking but, as shown here, even the dreariest can be transformed at dusk, with the lights of the city picking out landmarks not always visible during the day. To reduce internal reflections from the glass during these night 'flights', the lighting inside the capsules has been programmed to shift seamlessly from pure white, during embarkation and disembarkation, to dark blue as the capsules rise.

"I have been reading, recently, about the problems Gustave Eiffel encountered during the construction of his famous tower. There are so many parallels. I was particularly taken by his comment that, for him, the tower was a symbol of all the challenges that had been overcome — and, reading between the lines, you know he doesn't mean only the technical or engineering ones. For Julia and myself, it's the same with the wheel. There have been so many highs and lows to plough through over the past six years — to find the right partners, to convince the planners, to lock-in the funding and so on, a constant roller coaster — that you can't help but be reminded of that. But then there are other moments, when we catch sight of the wheel unexpectedly, when it sparkles and looks so good that we cannot help but feel proud of what we have achieved and to know that we were right to persevere."

David Marks
Director, Marks Barfield Architects

The wheel from Waterloo Bridge, reflected almost perfectly in the still water of a slack tide, early on a bright summer's morning. At the time of writing in March 2002, a little over two years after opening to the public, British Airways London Eye has already entertained well over seven million visitors, marking it as one of the most successful visitor attractions in Europe.

It is somewhat humbling to realise that the majority of the construction photographs illustrating this book were taken within a 12 month period — between March 1999 and February 2000. The work that was achieved in that time, in its range, its scale and complexity, was truly remarkable and it was a privilege to share, if only in part, in the commitment and enthusiasm of those involved in the successful completion of this extraordinary project.

A great many people helped me along the way during that time, either by sorting out the practical issues of access and travel — and so ensuring I was, more often than not, in the right place at the right time — or, and more importantly, by finding the time in their busy schedules to talk to me and explain what was going on. There are far too many to name here, but to one and all I extend my sincere thanks. I can only hope I have captured something of your passion and integrity in this book.

On a more practical note, for all their help during the design and production of this book I would also like to extend my thanks to the following: Frank Anatole, Annie Booth, Julia Barfield, Jane Capper, Lesley Chisholm, Sandra Colosimo, Andrea Conti, Roberto Conti, Christianne and Robert Long, David Marks, William McElhinney, Prue Saunders, Renzo Toti and his team in the pre-press department at Conti Tipocolor, and Nick Wood for his generous help with the photography.

Please note that the photographs taken by Ian Lambot now form part of a permanent archive held and supervised by the British Airways London Eye Company.

Ian Lambot, March 2002

Rather than identify each picture individually, the figure in brackets below indicates only the number of images attributed to each contributor on that specific page, or spread, where there is more than one photograph.

Nic Bailey 152/153(1) 153(2)
The Booth Family Collection 97(1)
FAG Kugelfischer Georg Schäfer AG 60(1) 61(1) 62/63 63(2)
Marks Barfield Architects 8 11 12(2) 14(4) 14/15 18 19 20/21 74(3) 153(1)
Ian Lambot 21(2) 22 24(3) 25(2) 26(2) 26/27 27(2) 28(4) 29(7) 30(4) 30/31 36(3) 36/37 38 38/39 40(2) 40/41(1) 42(2) 42/43 44 45 46(4) 50(2) 50/51 52 53(1) 54(3) 55(2) 56 56/57 58(2) 58/59 60(3) 61(6) 64(3) 64/65 68 68/69 69(2) 70/71 71 72 78(3) 79(2) 80(2) 81(4) 82 82/83 86(1) 86/87 88(2) 90/91 92(2) 94(2) 94/95 95 96/97 97(1) 98(2) 98/99 99 104/105 105(2) 106(5) 107(3) 108/109(2) 109(3) 110(3) 111(4) 112(3) 112/113 114 115 116 116/117 117(2) 118(5) 119(5) 120(4) 122(7) 123(4) 126/127(1) 127(2) 128(4) 129(7) 130(2) 130/131 132(1) 132/133(2) 134/135 136/37 138(2) 138/139 139(2) 140/141(2) 141(3) 142(4) 143(7) 144/145 145 149 150 152/153(1) 154/155(2) 156(4) 157(3) 158 158/159 160(6) 161(4) 162/163(2) 163 164(5) 165(4) 167(3) 168(5) 169(1) 172(3) 173(5) 174(5) 175(5) 177 178(7) 179(4) 180/181 181 182(1) 183(4) 184/185 186(4) 187(7) 188(4) 189 190/191 191 192(3) 193 194(3) 194/195(1) 199 202
Museum of London 10/11
Nick Wood Cover 2/3 16/17 17 21(1) 23 24(2) 25(4) 32/33/34 35 40(1) 40/41(1) 47 48 48/49 49 53(1) 54(1) 64(1) 66(2) 67 73 74/75 76(2) 76/77 77 79(2) 80(1) 81(2) 84 85 86(2) 88/89 91(2) 92(1) 93 100/101/102 103 107(1) 121 124/125 126/127(1) 127(1) 132(1) 134(2) 135 137(2) 146/147/148 151 154(3) 166/167 168(1) 168/169 169(1) 170(3) 170/171(2) 172(1) 176 182(3) 185 194/195(1) 196/197/198 200 201
Nick Wood / Marks Barfield Architects 9 13

Credits